TRENDS IN APPLIED THEORETICAL CHEMISTRY

TOPICS IN
MOLECULAR ORGANIZATION AND ENGINEERING

Volume 9

Honorary Chief Editor:

W. N. LIPSCOMB (*Harvard, U.S.A.*)

Executive Editor:

Jean MARUANI (*Paris, France*)

Editorial Board:

The titles published in this series are listed at the end of this volume.

Trends in Applied Theoretical Chemistry

Edited by

LUIS A. MONTERO
Department of Chemistry, University of Havana, Havana, Cuba

and

YVES G. SMEYERS
CSIC, Madrid, Spain

Springer Science+Business Media, B.V.

Library of Congress Cataloging-in-Publication Data

```
Trends in applied theoretical chemistry / edited by Luis A. Montero
  and Yves G. Smeyers.
      p.   cm. -- (Topics in molecular organization and engineering ;
  v. 9)
    Includes index.
    ISBN 978-94-010-5102-6      ISBN 978-94-011-2498-0 (eBook)
    DOI 10.1007/978-94-011-2498-0
    1. Chemistry, Physical and theoretical--Congresses.   I. Montero,
  Luis A.  II. Smeyers, Yves G.  III. Series.
  QD450.T74   1992
  541.2--dc20                                                92-12289
```

ISBN 978-94-010-5102-6

Printed on acid-free paper

"The logo on the front cover represents the generative hyperstructure of alkanes", printed with permission from J.E. Dubois, Institut de Topologie et de Dynamique des Systèmes, Paris, France.

TABLE OF CONTENTS

PART IV: MOLECULAR SPECTROSCOPY

INTRODUCTION TO THE SERIES

The Series 'Topics in Molecular Organization and Engineering' was initiated by the Symposium 'Molecules in Physics, Chemistry, and Biology', which was held in Paris in 1986. Appropriately dedicated to Professor Raymond Daudel, the symposium was both broad in its scope and penetrating in its detail. The sections of the symposium were: 1. The Concept of a Molecule, 2. Statics and Dynamics of Isolated Molecules; 3. Molecular Interactions, Aggregates and Materials; 4. Molecules in the Biological Sciences, and 5. Molecules in Neurobiology and Sociobiology. There were invited lectures, poster sessions and, at the end, a wide-ranging general discussion, appropriate to Professor Daudel's long and distinguished career in science and his interests in philosophy and the arts.

These proceedings have been arranged into eighteen chapters which make up the first four volumes of this series: Volume I, 'General Introduction to Molecular Sciences'; Volume II, 'Physical Aspects of Molecular Systems'; Volume III, 'Electronic Structure and Chemical Reactivity'; and Volume IV, 'Molecular Phenomena in Biological Sciences'. The molecular concept includes the logical basis for geometrical and electronic structures, thermodynamic and kinetic properties, states of aggregation, physical and chemical transformations, specificity of biologically important interactions, and experimental and theoretical methods for studies of these properties. The scientific subjects range therefore through the fundamentals of physics, solid-state properties, all branches of chemistry, biochemistry, and molecular biology. In some of the essays, the authors consider relationships to more philosophic or artistic matters.

In Science, every concept, question, conclusion, experimental result, method, theory or relationship is always open to reexamination. Molecules do exist! Nevertheless, there are serious questions about precise definition. Some of these questions lie at the foundations of modern physics, and some involve states of aggregation or extreme conditions such as intense radiation fields or the region of the continuum. There are some molecular properties that are definable only within limits, for example, the geometrical structure of non-rigid molecules, properties consistent with the uncertainty principle, or those limited by the neglect of quantum-field, relativistic or other effects. And there are properties which depend specifically on a state of aggregation, such as superconductivity, fer-roelectric (and anti), ferromagnetic (and anti), superfluidity, excitions, polarons, etc. Thus, any molecular definition may need to be extended in a more complex situation.

Chemistry, more than any other science, creates most of its new materials. At least so far, synthesis of new molecules is not represented in this series, although the principles of chemical reactivity and the statistical mechanical aspects are included. Similarly, it is the more physico-chemical aspects of biochemistry, molecular biology and biology itself that are addressed by the examination of questions related to molecular recognition, im-munological specificity, molecular pathology, photochemical effects, and molecular communication within the living organism.

Many of these questions, and others, are to be considered in the Series 'Topics in Molecular Organization and Engineering'. In the first four volumes a central core is presented, partly with some emphasis on Theoretical and Physical Chemistry. In later volumes, sets of related papers as well as single monographs are to be expected; these may arise from proceedings of symposia, invitations for papers on specific topics, initiatives from authors, or translations. Given the very rapid development of the scope of molecular sciences, both within disciplines and across disciplinary lines, it will be interesting to see how the topics of later volumes of this series expand our knowledge and ideas.

WILLIAM N. LIPSCOMB

PREFACE

The present volume gathers a series of selected and updated contributions presented at the International Symposium on Applied Theoretical Chemistry held in Havana, Cuba, July 2–6, 1990. This Symposium was intended to illustrate current applications of Theoretical Chemistry in different fields of Physical Chemistry.

Theoretical Chemistry has become a powerful tool of investigation in all areas of Chemistry, Biochemistry, and Physical Chemistry. The plenary lectures given in the Symposium were classified into four topics: Atom-Surface Interactions, Chemical Reaction Mechanisms, Molecular Structure and Properties, and Molecular Spectroscopy. We retain the same division in this volume.

Over 60 scientists from Cuba, Finland, France, Germany, Great-Britain, Hungary, Italy, Spain, Sweden, USA, USSR, and Venezuela participated in the Conference. Twenty plenary lectures were given by distinguished members of the international scientific community. Furthermore, a large number of posters were presented by younger experts in various fields of Theoretical Chemistry.

This International Symposium was organized by the Faculty of Chemistry of the University of Havana and the Cuban Chemical Society. It was an opportunity to bring together in Havana several outstanding scientists from various countries of the world. Havana is worldwide renown for its wonderful climate, the hospitality of its inhabitants, and the proximity of beautiful touring resorts.

We acknowledge the support of the University of Havana, the Cuban Academy of Sciences and UNESCO to help organize this Conference. Many students, postgraduates and colleagues related to the Laboratory of Computational and Theoretical Chemistry, and to the Faculty of Pharmacy of the University of Havana, especially Dr. J.R. Alvarez-Idaboy and Mr. C. Navajas, were key persons in managing this meeting. The Higher Pedagogical Institutes of Pinar del Rio and Havana contributed with resources and the valuable help of Mrs. R. Gonzales-Jontes and L.R. Sierra-Medina. The board of the Dean of the Faculties of Biology and Chemistry, the heading bodies of the Institute of Materials and Chemicals, and the Office of International Affairs of the University of Havana also gave a very valuable support. Many other individuals contributed decisively to overcome occasional difficulties and make successful that important meeting for international friendship and scientific exchange.

LUIS A. MONTERO (Havana, Cuba)
YVES G. SMEYERS (Madrid, Spain)

CHEMISORPTION ON METAL SURFACES. A CLUSTER MODEL APPROACH

ULF WAHLGREN and PER SIEGBAHN
Institute of Physics
University of Stockholm
Vanadisvägen 9
S-113 46 Stockholm
Sweden

ABSTRACT. In this contribution we describe a cluster model developed for investigations of reactions on transition metal surfaces. The model is based on a one-electron ECP description of the metal atoms. *Ab initio* calculations on O and H chemisorbed on five atom copper and nickel clusters are presented and compared with ECP results. Cluster size effects are investigated by calculations on several adsorbates using clusters with up to 50 atoms. A state selection criterion which allows quite small clusters to be used as models of an infinite surface is described.

1. Introduction

Ab initio quantum chemistry has undergone a rapid development the last twenty years. At the end of the 1960's high accuracy quantum chemical results for molecules were essentially restricted to the case of the hydrogen molecule and some other few-electron diatomic molecules. Systems of chemical interest were only treated at a low accuracy semi-empirical level. At present, gas phase reactions can be routinely handled for up to 10 (and more) first row atoms to a very high degree of accuracy [1]. *Ab initio* quantum chemical calculations have therefore more and more become a natural ingredient in studies in most areas of modern chemistry. The goal of the project covered in this chapter is to carry this potentially highly accurate approach over to the area of surface science.

From a molecular chemists point of view the simplest and most natural model of a surface is a cluster of metal atoms. One can today perform medium level accuracy (Hartree-Fock) calculations with up to 5–10 transition metal atoms using conventional SCF (Self Consistent Field) methods. Using the direct SCF approach [2] one has been able to treat up to 34 copper atoms, but these calculations are quite expensive and will in the near future mainly be interesting as bench mark type calculations. If electron correlation of the $3d$-electrons is to be included a calculation including 4-5 transition metal atoms is today still a challenge. It is clear that these clusters are too small to be directly useful as models of surfaces, and simplifying assumptions must be made. The cluster model which is described in the present work is based on the Effective Core Potential approximation, and it involves a parametrization of the metal core including the d-electrons.

L. A. Montero and Y. G. Smeyers (eds.), Trends in Applied Theoretical Chemistry, 1–17, 1992.

The pioneering work on cluster models in connection with conventional quantum chemical methods, was made by Goddard and coworkers [3–5]. They showed that quite reasonable results can be obtained if the transition metal atoms are treated as one-electron systems using Effective Core Potentials (ECP's). This is of course a dramatic simplification and makes modelling of surfaces with up to a hundred atoms feasible. An important question is of course how much accuracy is lost by this one-electron ECP approximation.

In the present article we will describe the Effective Core Potential (ECP) model and the one-electron version of this approximation where the metal atoms are described as pseudo one-electron atoms (i.e. all remaining electrons including the d electrons are regarded as core). We will also briefly describe the Core Polarization Potential (CPP) model, which is a semi-classical model describing the effects of the most important part of the correlation involving the d-electrons.

Apart from the accuracy of the one-electron ECP approximation, the most fundamental question in the cluster modelling of chemisorption energetics is how many atoms are needed to reproduce actual metal surface conditions. It is often taken for granted that the Fermi level has to be well reproduced by a cluster if it should be at all reasonable as a surface model. Similar opinions exist about the reproduction of the continuous density of states. A quantitative reproduction of the Fermi level, or the work function, requires very large clusters. One can show [6] that a spherical cluster with radius 10 Å (containing up to 400 atoms) will have an ionization potential more than half an eV too high compared to a real surface. A basic investigation of the surface project was to systematically calculate the chemisorption energy for clusters of different sizes in order to try to understand the origin of the cluster oscillations in detail. It turns out that, by using a molecularly oriented picture rather than the conventional solid state oriented picture, it was indeed possible to understand the origin of the cluster oscillations [7]. In this new picture the exact position of the Fermi level is of a rather limited importance for the strength of the chemisorption bond [8] and, more importantly, with this new understanding it becomes possible to correct the cluster chemisorption energies to be useful for surface predictions.

We will in this work describe the results of a thorough investigation of the atomic chemisorption of hydrogen, oxygen, fluorine and CH_2 on nickel surfaces, and explain the role of the electron structure in the chemisorption process.

2. Theory

In this section we will describe the ECP method, the role of the d-electrons in the chemisorption process, the one electron ECP approximation and also the simplified Core Polarization Potential method to calculate the influence of the d-correlation on the chemisorption properties.

2.1. THE ECP METHOD

The ECP method dates back to 1960, when Phillips and Kleinman suggested an approximation scheme for discarding core orbitals in band calculations [9]. The idea was to modify the Fock operator such as to make the core and the valence orbitals degenerate:

$$\hat{F} \rightarrow \hat{F} + \sum_c (\epsilon_v - \epsilon_c) \, | \, c > < c \, |$$

The solutions corresponding to the core and the valence orbitals could then be rotated in such a way as to remove the nodes in the valence orbital while keeping its valence character in the outer atomic regions. This method was called the Pseudopotential method by Phillips and Kleinman. Only one single valence electron was considered in the original pseudopotential method.

The generalization of the pseudopotential method to molecules was done by Bonifacic and Huzinaga [10] and by Goddard, Melius and Kahn [11] some ten years after Phillips and Kleinman's original proposal. In the molecular pseudopotential or Effective Core Potential (ECP) method the core-valence interactions are approximated with l dependent projection operators and a totally symmetric screening type potential. The new operators, which are parameterized such that the ECP operator should reproduce atomic all-electron results, are added to the Hamiltonian and the ECP equations are obtained variationally in the same way as the usual Hartree Fock equations.

The ECP method which will be discussed henceforth is derived from Huzinaga and Bonifacic. It should be mentioned that in our model the full nodal structure of the valence orbitals is always kept. In our early ECP application on first row transition metals the only orbitals which were included were $3d$ and $4s$ [12–14]. However, experience showed that in certain cases it was important also to include (frozen) $3s$ and the $3p$ orbitals in the valence space [13, 15], and ECP's with these characteristics were accordingly developed [16].

The equations used in the frozen ECP formalism are as follows:

$$\hat{F} = \hat{h}^{eff} + \sum_{c}^{val}(2\hat{J}_i - \hat{K}_i)$$

where

$$\hat{h}^{eff} = \hat{T} + \hat{V}^{eff} + \hat{P} + \sum_{j}^{frozen}(2\hat{J}_j - \hat{K}_j)$$

$$\hat{V}^{eff} = \left(\frac{-Z^{eff}}{r}\right)(1 + M_1 + M_2)$$

$$M_1 = \sum_{p} A_p exp(-\alpha_p r^2)$$

$$M_2 = \sum_{q} r C_q exp(-\gamma_p r^2)$$

$$\hat{P} = \sum_{k} |\phi_k > B_k < \phi_k|$$

The parameters which enter these equations are B_k, A_p, α_p, C_q and γ_q. The B_k values are usually chosen as the absolute value of the corresponding core orbital energies. A_p, α_p, C_q and γ_q are calibrated in atomic calculations such as to reproduce the orbital energies and shapes of the valence orbitals. The outer core orbitals are expressed in the valence basis set by a least squares fitting procedure and are kept frozen during both the parameter fitting and the molecular calculations. This ECP method been shown to give quantitatively correct results for many sysyems, including transition metal complexes.

As a consequence of the development of both software and hardware the traditional ECP method has became rather obsolete for the first row transition metals. An obvious case where the method still is important is of course for heavier elements, where relativistic effects can be parametrized into the ECP:s. This method has been successfully applied by many workers for the second and third transition metal series.

A different use of the ECP formalism was developed with the emerging interest in reactions on metal surfaces. With present day computers it is not realistic to carry out quantitative *ab initio* calculations on more than 5-10 metal atoms. However, if the troublesome d shells can be approximated, much larger clusters can be handled. Melius *et al.* [3–5] were the first to suggest a computational scheme in which the d electrons are also included in an ECP. This type of ECP, in which only the $4s$ electrons are explicitly accounted for in the calculations, was used by Goddard *et al.* [11] in calculations of oxygen and hydrogen chemisorption on nickel surfaces. These and similar one-electron ECP's have also been used successfully by Bagus *et al.*, by Bauschlicher and by Siegbahn *et al.* in applications where one metal atom is described at the all-electron level and the one-electron ECP atoms are used to mimic a metallic surrounding [17–19].

There are several prerequisites which have to be fulfilled for the one electron ECP approach to be applicable. In the case of metal clusters the atomic configuration must be known, i.e. one must safely be able to assume a $d^n s^2$ or a $d^{n+1} s^1$ configuration on the atoms in the cluster. The d orbitals should not form covalent bonds neither within the cluster nor between the cluster and the adsorbate. Ferromagnetic metals and copper are likely to have these properties, but for other metals, particularly to the left in the periodic table and in the second and third transition series, this is not so clear.

Since the d orbitals are not allowed to relax in a one-electron ECP it may appear that a third prerequisite is that the frozen d approximation should be valid, i.e. the relaxation of the d orbitals should not influence the bonding appreciably. In fact the effect of d orbital relaxation on chemisorption energies is appreciable, and this assumption is thus not fulfilled. We will adress the problem concerning the role of the d-orbitals below and explain why it is nevertheless possible to develop a cluster model based on a one-electron ECP.

2.1.1. The role of the d-orbitals.   The d orbital relaxation energy is defined as the difference at the SCF level between the chemisorption energy obtained at the all-electron SCF level when the d orbitals are frozen in their atomic shapes and when they are allowed to relax in the molecular surrounding. The metals selected for the investigation of the role of the d orbitals were nickel and copper. The ferromagnetism of nickel makes it likely that the d orbitals have an essentially atomic character in the metal, and for both nickel and copper it is reasonable to assume that the d orbitals are largely inert during surface reactions. The cluster used was a five-atom cluster, arranged in a square pyramid with an oxygen atom approaching the base of the pyramid. This system serves as a model for oxygen chemisorption at a fourfold hollow site on a (100) surface of the metal.

The results of the all-electron investigations were quite surprising, particularly for the nickel system. The d shell relaxation energy, i.e. the difference obtained for the oxygen binding energy with frozen and relaxed d orbitals, was 44 kcal/mol for Ni_5O and 17 kcal/mol for Cu_5O [20, 21]. The total oxygen binding energy in Ni_5O is 42 kcal/mol at the SCF level, which means in a sense that all the binding at this level comes from the d orbital relaxation. As will become clear below, the qualitative aspects of the bond formation are

TABLE I
CSOV analysis of the bonding in Cu_5O. ΔE is the difference from a relaxed $3d$ calculation. The chemisorption energy D_e is calculated relative to $Cu_5(^4A_2) + O(^3P)$. All energies in kcal/mol.

Step	ΔE	D_e	$3d$-pop	Q(O)
Step 1	31.8	7.6	50.00	−0.85
Step 2	1.6	25.4	49.78	−0.88
Step 3	0.0	27.0	49.76	−0.84

correctly described already at the frozen $3d$ level, however.

There are two possible explanations of the large effect of the d orbital relaxation on the oxygen binding energy. The system may be trying to minimize the repulsion between the adsorbate and the d orbitals, in which case the origin of the relaxation is a polarization or hybridization of the d orbitals away from the adsorbate. This picture is consistent with the larger relaxation obtained for Ni than for Cu, since Ni_5 has the possibility to minimize the repulsion by rotating the open and the closed d orbitals. The d orbitals may of course also bind covalently to the adsorbate. This possibility seemed initially not very likely since a localization of the d orbitals in the relaxed case yielded quite pure d orbitals. However, orbital coefficients may not be a certain measure of covalency effects, and in order to quantify the covalent contribution of the d orbitals to the cluster adsorbate bond a CSOV [22] analysis was carried out on Cu_5O [23] (the presence of open d orbitals on Ni_5O makes a corresponding analysis more difficult).

The Constrained Space Orbital Variation, or CSOV technique is a method for analysing in detail different contributions to a chemical bond [22]. In the CSOV method, subsets of the occupied orbitals are frozen or otherwise constrained at different stages of the calculation, sometimes in combination with constraints on the virtual space. Our CSOV analysis consisted essentially of three steps; a first step where the binding energy was calculated with all d orbitals frozen in their atomic shapes, a second step in which the d orbitals were relaxed but where the rest of the valence space (as obtained from step 1) was frozen , and finally a fully relaxed calculation. Since the bonds between the valence orbitals of the cluster and the adsorbate are formed in the first step and not allowed to change in the second step the energy difference between the second and the third steps gives a measure of the importance of the d orbitals for the covalent bonds. The result of the CSOV analysis are shown in Table I. The energy difference obtained between the second and third steps was only 2 kcal/mol, and the effect of the d orbitals on the binding is thus caused almost exclusively by polarization of the d orbitals in the cluster. The repulsion between the adsorbate and the d orbitals at the frozen orbital level can be rationalized in terms of Pauli exclusion effects, i.e. the orthogonalization of the adsorbate orbitals against the d orbitals in the cluster. We should mention here that in the second step described above we actually carried out several calculations using different subsets of the virtual space as well but, this gave no conflicting information.

A similar investigation on the Ni_5O [24] cluster gave a somewhat different result. In this case the increase in the oxygen binding energy obtained between the calculation where the d orbitals were relaxed in a frozen valence space and the fully relaxed calculation was 18 kcal/mol. However, the CSOV analysis becomes more complicated and actually somewhat ambiguous on metals with open d shells, largely because of Pauli repulsion effects. Our experience shows that the orbital configuration selected for the frozen orbital calculation does not change when the d orbitals are relaxed in the field of the frozen valence space chosen for the open d shells. The computed relaxation will thus depend on the d occupation used at the frozen level, and Pauli repulsion terms may become severely overestimated since even minute deformations in the outer regions of the d orbitals can involve rather large energy effects. In fact, the high localizability of the $3d$ orbitals makes a $3d$ covalency effect of 18 kcal/mol appear rather high. Considering the ambiguities in the CSOV analysis in the Ni_5O case we believe that 18 kcal/mol represents an overestimate of the $3d$ covalency effects in Ni_5O by maybe 8-10 kcal/mol . However, the results do indicate that the separability of the valence and the d orbitals is better for copper than for nickel which may be interpreted as a higher degree of d covalency in the adsorbate substrate bonding in Ni_5O than in Cu_5O. The total effect is at any rate rather small compared to the total binding energy of 115–130 kcal/mol.

All the above calculations were carried out using a fully contracted d orbital on the metal atoms. If a radial polarization of the d shell or if covalency is important one would expect the results to be sensitive to the basis set used to describe the d orbital. In order to investigate this effect a larger calculation using a triply split d shell was carried out on Ni_5O [25]. The chemisorption energy obtained with the two different basis sets was remarkably stable, 41.5 kcal/mol for the small basis and 45 kcal/mol with the large basis, a result which confirms the conclusion that d covalency is not important for the bond between a nickel surface and oxygen.

2.1.2. The one-electron ECP:s.   Actually, a small d orbital relaxation is not a necessary prerequisite for the development of a one-electron ECP. The essential point is that the relaxation effect is dominated by the cluster and not by the adsorbate, and this is likely to be the case since the relaxation is not dominated by covalency effects.

In the one-electron ECP the effects from all electrons except for the 4s are parametrized into the ECP operators. One evident consequence is that the atomic ground state can no longer be used to find all the parameters needed. However, the most important consequence of the approximation is that the d orbital projection operator will not play the role of a simple level shift operator any longer. On the contrary, it will contribute actively to the bonding by raising the energy of the system when an approaching orbital is starting to overlap with the d orbital, which of course is precisely the effect of a "real" d orbital as well.

The only way to determine the parameters of the d projection operators at the atomic level would be to consider excited states of the atom with an occupied $4d$ orbital. The sensitivity of the $4d$ orbital to the parameters of the d projection operator is, however, very small [26].

The problem thus arises how to determine the parameters of the d-type projection operator. The route choosen was to optimize the d orbital projection operator by comparing with molecular or cluster calculations at the all-electron level. In our one-electron ECP approach we have decided to make the parameterization of the d projection operator by comparing with all electron calculations on oxygen chemisorption at the hollow position on a five atom metal cluster [26]. The parameterization of a one-electron ECP describing

TABLE II

SCF results for hydrogen and oxygen on Ni_5O and Cu_5O. The all-electron r_e value for CU_5H is not geometry optimized.

System	Case	r_e	D_e	ω_e	O−pop
Ni_5O	All-electron	1.92	41.6	381	−1.04
Ni_5O	ECP	1.95	42.9	347	−1.04
Cu_5O	All-electron	1.71	27.3	328	−0.85
Cu_5O	ECP	1.66	27.6	316	−0.88
Ni_5H	All-electron	2.38	36.8	835	−0.63
Ni_5H	ECP	2.49	33.4	847	−0.51
Cu_5H	All-electron	(2.50)	25.3	–	−0.64
Cu_5H	ECP	2.51	24.5	764	−0.42

chemisorption on a cluster with d orbitals frozen in their atomic shapes yielded a B_k value about twice as large as the orbital energy. In the relaxed d orbital calculations it turned out to be impossible to obtain both the chemisorption energy and the distance above the surface with the original shape of the d orbital from the atomic calculation. This result is not so surprising, since a large fraction of the d relaxation energy is caused by a polarization of the d orbitals which of course affects their shape. The solution to this problem was to modify the form of the d projection operator to 'soften' it in the outer regions by introducing a second diffuse and slightly attractive d projection operator.

The exponents and the expansion coefficients in the basis set used with the ECP were determined by a least-squares fit to the all-electron orbitals. All the parameters entering the screening potential were determined by fitting the shape and the energy of the $4s$ orbital to the corresponding all-electron results. The B_k values entering all but the $3d$ type projection operators were chosen in the standard way, i.e. as the negative orbital energies of the corresponding core orbitals. A frozen $3s$ orbital was included in the ECP on all centres near the adsorbate.

The final calibration of the B_{3d} parameters were carried out on a five atom metal cluster with an adsorbed oxygen atom at the fourfold hollow position. The binding energies and the bond distances were reproduced to within 1 kcal/mol and 0.05 bohr for Ni_5O [26] and Cu_5O [21] (see Table II). The vibrational frequencies computed using the one electron ECP's are somewhat too low, by 40 cm^{-1} for Ni_5O and by 10 cm^{-1} for Cu_5O, indicating a somewhat underestimated repulsion at shorter bond distances.However, the geometrical configuration, with an oxygen atom sort of sliding in between four nickel atoms, will make the vibrational frequencies very sensitive even to minor errors in the interatomic potentials.

It is not evident that one-electron ECP's which have been optimized for oxygen can be used for other adsorbates. If e.g. the d orbitals bind covalently to oxygen to any appreciable extent, we would expect the model to describe adsorption of other adsorbates less satisfactorily. Calculations on Ni_5H and Cu_5H yielded, however, completely satisfactory results (see Table II). The difference between the ECP and the all-electron binding energies were 3 kcal/mol for Ni_5H and 1 kcal/mol for Cu_5H, and the distance above the surface was overestimated by 0.1 a.u. (for Ni_5H) . This is not much considering that the potential energy surface is very flat.

Adsorption of oxygen and hydrogen at hollow sites on Ni(100) and Cu(100) is described satisfactorily by one-electron ECP clusters. When adsorption at on-top positions are considered, effects directly involving the metal d shells become important. An example of this is the dissociation of molecular hydrogen at the on-top position on a Ni(100) surface where $s-d$ hybridization is very important both for intermediate states and for the barrier to dissociation. In such cases it is necessary to include one all-electron atom at the on-top site in the calculations. For reactions which take place at bridge sites it is desirable to use two all-electron atoms, but in this case the results from all ECP calculations can be corrected by comparing with calculations with an adsorbate and only two all-electron metal atoms.

2.2. THE CORE POLARIZATION POTENTIAL METHOD

The Core Polarization Potential (CPP) method [27] is based on a classical effective field operator where the core-valence correlation is viewed as resulting from the interaction between the field generated by the valence electrons and an induced core-dipole moment proportional to the core-polarizability α_c.

$$\hat{V}_{CPP} = -\frac{1}{2}\sum_c (\alpha_c \mathbf{f}_c^2 - 2 \cdot \mathbf{f}_c \cdot \mu_c^0 + \mathbf{f}_c^0 \cdot \mu_c^0)$$

The operator $\hat{V}_{CPP}$ acts on the valence electrons only and must thus be projected onto the valence space. The static polarization of the cores is included in the operator through the term $-\frac{1}{2}\sum_c \alpha_c \mathbf{f}_c^2$ and the terms containing the expectation value, μ_c^0, of the core dipole moment thus correct for the static core polarization actually obtained with the basis set used [27]. The fields, $\mathbf{f}_c$, contain contributions from both the surrounding cores and the valence charge density. In order to avoid unphysical contributions from penetration into the core region a cut-off function, $C(\rho_c, \mathbf{r}_{ci})$, is applied to the fields generated by the valence electrons

$$C(\rho_c, \mathbf{r}_{ci}) = (1 - exp(-(\frac{\mathbf{r}_{ci}}{\rho_c})^2))^2$$

where ρ_c is a cut-off parameter which is determined so that the experimental or core-valence CI ionization potential for metal atom is reproduced.

The CPP operator thus contains two basic parameters, the core dipole polarizability (α_c) and a cut-off radius (ρ_c) inside which the field generated by the valence electrons is made to disappear exponentially.

The core polarization potential (CPP) method has been shown to give very accurate results for the alkali metals [27, 28]. Pettersson and Åkeby have modified and extended the method for the more difficult copper compounds, and investigated the importance of core-valence polarization on the oxygen chemisorption energy on a Cu$_5$ cluster [29]. The calculations have been carried out at the all electron level using both a completely contracted and a triply split d shell on copper [29, 30]. Size-consistent CI calculations were also carried out using the correlated pair function (CPF) method [31] on Cu$_5$O correlating, the d electrons on the upper four copper atoms and the $2s$ and $2p$ shells on oxygen. The results of these calculations are shown in Table III.

TABLE III

Computed chemisorption energy for Cu_5O with and without explicit correlation of the $3d$ orbitals. The 3_d basis set has a triply split and 1_d basis set a totally contracted $3d$ orbital. The numbers within parenthesis are the number of correlated electrons. The 3_d results have been corrected for the basis set superposition error.

Basis	Method	r_e	D_e
3_d	SCF	1.75	27.1
3_d	CPF(11)	1.94	88.8
3_d	CPF(51)	1.71	102.0
1_d	CPF(11)	1.92	86.5
1_d	CPF(11)+CPP	1.53	99

3. The Cluster Model

The cluster model described in the present work is largely based on the one electron ECP approximation. It was originally designed as a model for chemisorption and reactions on metal surfaces, but lately a new application has become reactivities of small real metal clusters. The model is used as follows:

— At hollow positions on a metal surface all metal atoms are described by one electron ECP:s

— At on top positions the metal atom immediately underneath the adsorbate is described at the all-electron level. All remaining metal atoms are described by one electron ECP:s

— The valence correlation contribution to the substrate-surface bond is calculated using normal CI procedures.

— The core(3d)-valence correlation contribution to the bonding is, when included, obtained using the CPP method.

One of the central features of the model is that it is designed to mimick all electron calculations on small clusters, not primarily experimental results. This aspect is important, since all results can then be understood in terms of the basic computational models (independent particle model, correlation etc). Clearly a finite cluster is not equivalent to a full surface, and in order to analyze the model it is necessary to have all approximations used under control.

A central issue in the development of a surface cluster model is to understand the differences between a finite cluster and an infinite surface. In the early applications chemisorption energies showed erratic variations with the size of the cluster. This behavious of the cluster model could be understood from the electronic properties of the cluster. In a finite cluster there is a discrete set of excited levels, while for an infinite surface this discrete set of states is replaced by a continuous band. For a finite cluster there is thus a problem in just how to select the proper states to use for calculating e.g. chemisorption energies, since states which are excited relative to the ground state by perhaps 1-2 eV in the cluster becomes degenerate in the limit of an infinite surface. The concept of bond preparation makes it possible to select the relevant states for a small cluster such that the cluster model can be used as quantitative model for describing reactions on an infinite surface.

3.1. BOND-PREPARATION FOR HYDROGEN AND OTHER ADSORBATES

The simplest possible adsorbate for studying the cluster size dependence on the chemisorption energy is the hydrogen atom. In this case there is only one bonding mechanism, hydrogen is bound to the substrate with one covalent bond, and there are no other interactions that influence the chemisorption energy such as repulsion from closed shells on the adsorbate. Nevertheless, even for the hydrogen atom it is found that if the chemisorption energy is calculated as the difference between ground states at short and long distances between the adsorbate and the cluster (E_{Gr}), strongly oscillating results as a function of cluster size are obtained [7]. Of seven different clusters modelling the Ni(100) surface (Ni_5-Ni_{50}), it was found that in particular $Ni_{21}(12,9)$ gave a very poor chemisorption energy of 43.0 kcal/mol compared to the experimental value of 63 kcal/mol [32]. All the other six clusters gave chemisorption energies in the range 50-65 kcal/mol (these results include correlation effects). When the Ni(111) surface was modelled by ten different clusters (Ni_4-Ni_{40}) even more strongly oscillating results were obtained. This surface has the same experimental chemisorption energy of 63 kcal/mol, but the $Ni_{19}(12,7)$ cluster, for example, gave only a value of 27.5 kcal/mol and $Ni_{22}(12,7,3)$ gave 37.8 kcal/mol. It turns out that all these results can be understood in a rather simple way. The clusters that gave poor chemisorption energies all had only closed shells in symmetry A_1 (the same symmetry as the hydrogen $1s$ orbital), and also had large excitation energies to the first excited state which had a singly occupied a_1 orbital. However, if the chemisorption energies instead were calculated with respect to this excited state quite reasonable chemisorption energies were obtained for every cluster. This procedure was termed bond-preparation. The bond-prepared results for the seven clusters modelling the Ni(100) surface gave an average value of 60.3 kcal/mol with a standard deviation of only 3.0 kcal/mol, whereas for the ten clusters modelling the Ni(111) surface the corresponding results are 61.7 kcal/mol and 4.5 kcal/mol, respectively. The simplest way to understand what bond-preparation means is to interpret the chemisorption bond as covalent, thus requiring one singly occupied orbital both on the adsorbate and also locally on the substrate. For a real surface (unlike the situation for a finite cluster) this type of bond-prepared state should be accessible without any energy cost due to the band nature of the electronic states of an infinite surface. As will be seen below there is also another way to interpret the results for hydrogen, which turns out to be more applicable for the case of atomic oxygen chemisorption in particular.

The first adsorbate studied after the hydrogen atom was the methyl radical [33], and as already mentioned it behaved in essentially the same way as hydrogen. If the cluster was prepared with one singly occupied orbital of A_1 symmetry, stable chemisorption energies were obtained. Calculations of this type led to the presently probably best estimate for the chemisorption energy of methyl on Ni(111) of 50-55 kcal/mol. Previous estimates ranged from 30 kcal/mol to 70 kcal/mol. It should be added that accurate experimental determinations of the chemisorption energies for short lived adsorbates like methyl are extremely difficult to obtain.

The first more critical test case for the generality of bond-preparation was recently made for the case of atomic fluorine chemisorption on Ni(100). Since the concept of bond-

TABLE IV

Chemisorption of atomic fluorine in the 4-fold hollow site of Ni(100), energies in kcal/mol. E_{Gr} are the chemisorption energies calculated with respect to the ground states at short and long distances, whereas E_{Bp} are calculated relative to bond-prepared states.

Cluster	E_{Gr}	E_{Bp}
$Ni_5(4,1)$	121.7	121.7
$Ni_9(4,5)$	111.7	121.9
$Ni_{17}(12,5)$	99.3	117.9
$Ni_{21}(12,5,4)$	125.1	125.1
$Ni_{25}(12,9,4)$	120.4	120.4
$Ni_{29}(16,9,4)$	113.5	120.4
$Ni_{33}(12,9,12)$	128.1	128.1
$Ni_{37}(12,13,12)$	120.1	127.9
$Ni_{41}(16,9,16)$	119.2	122.8
$\overline{D}$	118.4	122.9
σ	7.7	3.3

preparation might appear to require the formation of a covalent bond, it is interesting to see what happens for an extremely ionic adsorbate. For the related system of F on Cu(100), the bonding has recently been analyzed in detail and it was concluded that the bonding is entirely ionic, and F on the Ni(100) surface should be similar [34]. One complication expected for an ionic adsorbate is that the ionization energy of the cluster could enter into the estimate of the chemisorption energy, and the ionization energy is known to converge very slowly with cluster size (as $1/R$, where R is the size of the cluster). This expectation turns out to be wrong, however. The results for the fluorine chemisorption energy for nine different clusters, calculated both with respect to ground states and with respect to bond-prepared states, are given in Table IV. Bond-preparation is done in exactly the same way as for the hydrogen atom. The most striking effect of bond-preparation is obtained for the $Ni_{17}(12,5)$ cluster, which has the poorest ground state chemisorption energy (E_{Gr}) of 99.3 kcal/mol but a bond-prepared chemisorption energy (E_{Bp}) of 117.9 kcal/mol in line with the results for the other clusters. Overall, the standard deviation is reduced from 7.7 kcal/mol down to 3.3 kcal/mol using bond-preparation.

As mentioned above, with a totally ionic bond it might be expected that the chemisorption energy should be sensitive to the ionization energy (the Fermi level) of the cluster. It is therefore rather surprising that no such sensitivity is found in the calculations. If one should try to explain the variation of the chemisorption energies in Table IV, which are successfully explained by the bond-preparation concept, by variation in IP's (ionization potentials) instead, there will be a total failure. For example, Ni_{17} has the smallest chemisorption energy at the SCF level of 75.3 kcal/mol but does not at all have the largest ionization energy. Using Koopmans' theorem an IP of 3.92 eV is obtained for ionizing an e-electron. Ni_{33} has the largest chemisorption energy of the clusters at the SCF level with 97.8 kcal/mol, but still has an IP of 4.69 eV which is thus as much as 18 kcal/mol larger than the one for Ni_{17}. The situation does not change much, even if only bond-prepared clusters and only ionization of an a_1 electron is considered. This is perhaps even more surprising since an a_1 electron has definitely moved from the cluster to fluorine. For example, the energy for ionizing an a_1 electron from Ni_{29} is 4.36 eV, while the corresponding IP of Ni_{37} is 5.06 eV, which is thus 16.1 kcal/mol larger. With the larger IP for Ni_{37} a smaller chemisorption energy would thus be expected for this cluster compared to Ni_{29}. The bond-prepared results show that the opposite is actually true, with a larger chemisorption energy for Ni_{37} (at the SCF level the difference is 9.9 kcal/mol). This contradiction is not a result of using Koopmans' theorem. Separate ΔSCF calculations for Ni_{29} and Ni_{37} lead to IP's of 4.07 eV and 4.47 eV, respectively, which still shows a large energy difference in the same direction as the Koopmans' values. The conclusion for fluorine chemisorption must be the same as the conclusion drawn earlier for hydrogen chemisorption [8], that the position of the Fermi level (lowest IP) is nearly irrelevant for describing chemisorption on clusters of these sizes.

In order to understand the results for fluorine, which is so ionic, an alternative formulation, which does not require a covalent bond formation, of how bond-preparation should be viewed is preferable. In this new picture, a reliable chemisorption energy is obtained if the electrons on the adsorbate fit into the electronic structure of the bare cluster. The practical consequence of this picture for hydrogen and fluorine is exactly the same as the bond-preparation described above. In the chemisorbed situation, essentially two electrons are located on hydrogen. With bond-preparation, these two electrons will occupy part of the space of a previously singly occupied cluster orbital. From the cluster point of view, no new orbitals are thus required to describe the chemisorption and a reliable chemisorption energy is obtained. The electrons on the adsorbate fit into the electronic structure of the bare cluster. If, on the other hand, the outermost cluster orbital with the same symmetry as the hydrogen orbital is doubly occupied, the addition of the hydrogen electron would mean that an additional, previously unoccupied, cluster orbital would have to be utilized for this electron, and a poor chemisorption energy is generally obtained. For fluorine the key orbital is the lone pair orbital pointing towards the cluster. For this orbital to fit into the electronic structure of the cluster, the outermost cluster orbital of the same symmetry has to be singly occupied, exactly as for hydrogen. With this new picture of the bonding, the insensitivity of the results on the IP of the cluster is expected. It is clearly in this case a poor picture, even for an ionic adsorbate, to view the bond formation in two steps with first the ionization of an electron from the cluster and then an addition of an electron on the adsorbate, since the electron basically stays in the same region of space when the bond is formed.

A few more points are worth mentioning concerning the results for fluorine. First, fluorine has one additional bonding mechanism compared to hydrogen, and this is with bond formation parallel with the surface in the E symmetry. This bonding mechanism was used by only one of the clusters, $Ni_{21}(12,9)$, and led to a chemisorption energy of 116.6 kcal/mol. With only one such example, it is at present not clear how this type of bonding should be compared to the bonding in the other clusters and this result was therefore not included in Table IV. Secondly, to reach a best prediction for the fluorine chemisorption energy a few corrections to the average value of 122.9 kcal/mol in Table IV has to be made. These correction are obtained from all-electron calculations on Ni_5F, from all-ECP calculations using a large basis set on fluorine (including up to two f-functions) and from counterpoise calculations for the basis set superposition error. Finally the $3d$ correlation effects were estimated based on CPP calculations and on all-electron correlated calculations on Cu_5F. The best estimate for the chemisorption energy is then obtained as 118.9 kcal/mol, which should be read as about 120 kcal/mol. There are no available experimental results to compare to. For further details of these calculations, see ref [35].

The description of the chemisorption bond, and thereby the concept of bond-preparation, is considerably more complicated for oxygen than for the adsorbates discussed above. The main reason for this is that not only one but two bonds are formed to the surface. To analyze the bonding for each cluster, the ground state occupation of the bare cluster should be subtracted from the ground state occupation for the cluster with oxygen adsorbed. When this is done three different bonding mechanisms can be identified. The first two of these are the obvious ones for the formation of two covalent bonds, with either two bonds parallel with the surface or with one bond parallel and one perpendicular to the surface. The third bonding mechanism is less obvious but turns out to be the one adopted by most clusters including the three largest clusters studied. For these clusters the occupations at short distance differ from the optimal bare cluster occupation only in the number of closed shells. The open shell occupations are the same. The closed shell occupations differ by two orbitals in symmetry A_1 and one orbital of E symmetry (occupied by four electrons). There is essentially only one way in which this bonding mechanism can be understood. The starting point is an oxygen atom which approaches the cluster with two electrons in the $2p_z$ lone pair (pointing down towards the surface) and with singly occupied $2p_x$ and $2p_y$ orbitals. As the chemisorption bond starts to form the $2p_z$ lone pair electrons will replace and thus kick out two electrons of the same symmetry (A_1) for the cluster. These two cluster electrons are moved over to the $2p_{x,y}$ orbitals of oxygen which are parallel to the surface and which originally have two holes. In this process the oxygen $2p$ electrons are partly delocalized over on the cluster, particularly for the $2p_{x,y}$ orbitals.

The above description of the main bonding mechanism for oxygen may seem quite different from the one found for fluorine, but it is in fact possible to describe also this latter bonding in a similar way. The starting point for fluorine is then with a doubly occupied $2p_z$ orbital pointing down towards the surface and a set of $2p_{x,y}$ orbitals with one hole. If the cluster has one singly occupied orbital of A_1 symmetry, which is the requirement for bond-preparation, the bonding proceeds with the $2p_z$ orbital replacing the singly occupied a_1 orbital and thus kicking out the electron from this orbital. This electron is then moved over to the hole in the $2p_{x,y}$ orbitals of fluorine, and the bonding is completed by delocalization of the $2p$ orbitals over on the cluster in complete analogy with the main bonding mechanism for oxygen (although the delocalization is less pronounced for fluorine than for oxygen). We note that the requirement of one singly occupied cluster a_1 orbital in the case of fluorine chemisorption follows from the fact that there is only one hole in the $2p_{x,y}$ orbitals on fluorine. With two holes in the $2p_{x,y}$ orbitals as on oxygen, there will be no such requirement on the cluster occupation in that case. It might further be added that with the above ionic description of the bonding mechanism, where electrons jump from the cluster to the adsorbate, the fact that electrons move between symmetries is not a significant aspect of the mechanism. In particular, for fluorine the orbital occupancy per symmetry can be kept by simply thinking of the fluorine $2p_z$ orbital as initially singly occupied.

With three possible bonding mechanisms available for oxygen and a main bonding scheme which does not lead to any requirements on the occupation of the cluster, it is perhaps not too surprising that already the ground state chemisorption energies E_{Gr} are reasonably stable from cluster to cluster, see Table V. There are, however, still two clusters

TABLE V

Chemisorption of atomic oxygen in the 4-fold hollow site of Ni(100), energies in kcal/mol. E_{Gr} are the chemisorption energies calculated with respect to the ground states at short and long distances, whereas E_{Bp} are calculated relative bond-prepared states.

Cluster	E_{Gr}	E_{Bp}
$Ni_5(4,1)$	110.6	110.6
$Ni_9(4,5)$	104.7	104.7
$Ni_{17}(12,5)$	106.3	106.3
$Ni_{21a}(12,9)$	101.8	101.3
$Ni_{21b}(12,5,4)$	112.6	112.6
$Ni_{25}(12,9,4)$	112.9	112.9
$Ni_{29}(16,9,4)$	98.6	105.4
$Ni_{33}(12,9,12)$	113.2	113.2
$Ni_{37}(12,13,12)$	114.1	114.1
$Ni_{41}(16,9,16)$	111.1	111.1
$\bar{D}$	108.6	109.2
σ	5.1	4.2
Exp.		115-130

modelling the Ni(100) surface, Ni_{21a} and Ni_{29}, for which no reasonable bonding scheme can be attributed. It is gratifying to note that precisely these clusters are the ones which give the poorest chemisorption energies. The assignment of bonding schemes has thus allowed us to identify the clusters for which the results should be most questionable. Another consequence of the large flexibility in selecting bonding scheme is that bond-preparation by going to an excited state of the cluster is less successful for oxygen than it has been for hydrogen and fluorine. For a cluster which is not bond-prepared, this flexibility often leads to an adjustment of the wave-function which tends to give quite reasonable chemisorption energies also in these cases. Bond-preparation is thus barely capable of moving the result for Ni_{29} into the range covered by the other clusters and the result for Ni_{21a} hardly changes.

Exactly as discussed for fluorine above, an improved prediction of the chemisorption energy for oxygen on Ni(100) can be obtained by doing additional calculations. When this is done a value of 130.0 kcal/mol is obtained, which should be read as about 130 kcal/mol. This is in perfect agreement with the most reliable experimental value [36] but differs from earlier calorimetric measurements giving 115 kcal/mol [37]. More than half of the increase from the average value 109.2 kcal/mol in Table V to the final estimate of 130 kcal/mol comes from $3d$ correlation effects (13 kcal/mol) and most of the other half from increasing the basis set on oxygen (8.0 kcal/mol) to include more d- and also f-functions.

Similar calculations as described above for the Ni(100) surface have also been performed for the Ni(111) surface. Just as for hydrogen chemisorption, modelling the Ni(111) surface by different clusters leads to somewhat larger oscillations of the chemisorption energies than for the Ni(100) surface. This appears to be related to a more complicated nodal pattern of the orbitals on the Ni(111) than on the Ni(100) surface, which can be seen

TABLE VI

Chemisorption of methylene in the 4-fold hollow site of Ni(100), energies in kcal/mol. E_{Gr} are the chemisorption energies calculated with respect to the ground states at short and long distances, whereas E_{Bp} are calculated relative bond-prepared states.

Cluster	E_{Gr}	E_{Bp}
$Ni_5(4,1)$	79.3	79.3
$Ni_9(4,5)$	66.7	66.7
$Ni_{17}(12,5)$	72.7	75.6
$Ni_{21a}(12,9)$	70.9	72.0
$Ni_{21b}(12,5,4)$	80.3	80.3
$Ni_{25}(12,9,4)$	79.9	79.9
$Ni_{29}(16,9,4)$	73.4	79.2
$Ni_{33}(12,9,12)$	78.3	78.3
$Ni_{37}(12,13,12)$	80.6	80.6
$Ni_{41}(16,9,16)$	80.3	80.3
$\bar{D}$	76.2	77.2
σ	4.7	4.3
Exp.		70-100

by studying the orbital contour plots. The final best prediction of the oxygen chemisorption energy in the three-fold hollow site of Ni(111) is 115 kcal/mol. There are no available experimental measurements for this energy but the result agrees well with expectations based on other similar results [38]. More details of the oxygen calculations can be found in ref [39].

The final adsorbate discussed in this subsection is methylene (CH_2). The results for the ground state and the bond prepared chemisorption energies of 10 different clusters modelling the Ni(100) surface are given in Table VI. Methylene was in all these calculations oriented in a plane perpendicular to the metal surface. The main difference in the chemisorption of oxygen and methylene is that oxygen has three bonding mechanisms but methylene only two. The bonding mechanism missing for methylene is obviously the one with two bonds parallel to the surface which is not possible with the present orientation of the molecule. Just as for oxygen, the E_{Gr} results are already quite stable for methylene, and in the cases where bond preparation is required the effect is only minor. The major reason for this is that the main bonding mechanism (the same as for oxygen) does not set any requirements on the occupation of the bare cluster in contrast to the case for hydrogen and fluorine. The final best estimate for the chemisorption energy of methylene is 99.2 kcal/mol, which should be read as about 100 kcal/mol. No direct measurements of the chemisorption energy of this short-lived radical exist but previous estimates range from 70 to 100 kcal/mol. The main corrections to the average value of 77.2 kcal/mol from Table VI comes from the inclusion of $3d$ correlation effects (9.7 kcal/mol) and the increase of the adsorbate basis set (6.5 kcal/mol), similar to oxygen, but for methylene there are also some minor (4.7 kcal/mol) multi-reference CI effects.

4. Conclusions

In this chapter we have summarized the present status of cluster modelling of chemisorption energetics, and shown that the cluster model can indeed be used for quantitative predictions of chemisorption energies. The key result in this respect is that, at least for strongly bound adsorbates, the cluster size oscillations of the chemisorption energy is easy to understand and also to correct for. The use of bond-prepared clusters leads in most cases to stable chemisorption energies even for rather small clusters of 10-20 atoms. Furthermore bond-prepared chemisorption energies are not only stable but also in quite good agreement with surface experiments. This means that the surface chemical bond is not as delocalized as had often been believed when, for example, embedding theories or infinite models were considered necessary. On the contrary, the surface bond must be regarded as quite localized, which is of course of utmost importance when small clusters are used as models. Finally, model calculations have shown that the position of the Fermi level is not as significant as results using other methods had indicated [6–10]. In fact, shifting the Fermi level by 7 eV by adding and removing electrons from a cluster, changed the hydrogen chemisorption energy by less than 0.1 eV.

References

[1] Hehre, W.R., Radom, L., Schleyer, P.R., and Pople, J.A.: 1986, *Ab Initio Molecular Orbital Theory*, Wiley, New York.

[2] Almöf, J., Faegri, Jr, K., and Korsell, K.: 1982, 'Principles for a Direct SCF approach to LCAO-MO *ab Initio* calculations', *J. Comput. Chem.* **3**, 385; Almöf, J. and Gropen, O., private communication.

[3] Melius, C.F., Upton, T.H., and Goddard, W.A.: 1978, 'Electronic properties of metal clusters (Ni_{13} to Ni_{87}) and implications for chemisorption', *Solid State Comm.* **28**, 501.

[4] Upton, T.H. and Goddard, W.A.: 1981, 'Chemisorption of H, Cl, Na, O and S on Ni(100) surfaces: A theoretical study using Ni_{20} clusters', in *CRC Critical Reviews, Solid State and Materials Sciences*, CRC Press, Boca Raton, p. 261.

[5] Upton, T.H. and Goddard, W.A.: 1979, 'Chemisorption of atomic hydrogen on large scale nickel cluster surfaces', *Phys. Rev. Lett.* **42**, 472.

[6] Wood, D.M.: 1981, 'Classical size dependence of the work function of small metallic spheres', *Phys. Rev. Lett.* **46**, 749.

[7] Panas, I., J. Schüle, P. Siegbahn, and U. Wahlgren: 1988, 'On the Cluster Convergence of Chemisorption Energies', *Chem. Phys. Lett.* **149**, 265.

[8] Panas, I. and Siegbahn, P.E.M.: 1990, 'The Nature of the Surface Chemical Bond – a Comparison Between the Molecular and Solid State Pictures', *J. Chem. Phys.* **92**, 4625.

[9] Phillips, J.C. and Kleinman, L.: 1959, 'A New method for calculating wave functions in crystals and molecules', *Phys. Rev.* **116**, 287.

[10] Bonifacic, V. and Huzinaga, S.: 1974, 'Atomic and molecular calculations with the model potential method', *J. Chem. Phys.* **60**, 2779.

[11] Melius, C.F. and Goddard, W.A.: 1974, '*Ab initio* effective potentials for use in molecular quantum mechanics', *Phys. Rev.* **A10**, 1528.

[12] Wahlgren, U.: 1978, 'Pseudopotential Calculations on some first, second and third row molecules. A comparative study', *Chem. Phys.* **32**, 215.

[13] Gropen, O., Wahlgren, U., and Pettersson, L.G.M.: 1982, 'Effective core potential calculations on small molecules containing transition metal ions', *Chem. Phys.* **66**, 459.

[14] Gropen, O., Wahlgren, U., and Pettersson, L.G.M.: 1982, 'Effective core potential calculations on the NiH_4^{2-} ion as a test case for studying rotational barriers', *Chem. Phys.* **66**, 453.

[15] Pettersson, L.G.M. and Strömberg, A.: 1983, 'A study of the value of the interaction integrals in effective core potential applications', *Chem. Phys. Lett.* **99**, 122.

[16] Pettersson, L.G.M., Wahlgren, U., and Gropen, O.: 1983, 'Effective core potential calculations using frozen orbitals. Applications to transition metals', *Chem. Phys.* **80**, 7.

[17] Bagus, P.S., Bauschlicher, C.W., Nelin, C.J., and Laskowski, B.C.: 1984, 'A proposal for the proper use of pseudopotentials in molecular orbital cluster model studies of chemisorption', *J. Chem. Phys.* **81**, 3594.

[18] Bauschlicher, C.W.: 1986, 'Coverage dependent effects on metal surfaces: O, S, F and Cl on Ni', *J. Chem. Phys.* **84**, 250.

[19] Siegbahn, P.E.M., Blomberg, M.R.A., and Bauschlicher, C.W.: 1984, 'The dissociation of H_2 on the Ni(100) surface', *J. Chem. Phys.* **81**, 2103.

[20] Panas, I., Siegbahn, P., and Wahlgren, U.: 1988, 'Model studies of the chemisorption of hydrogen and oxygen on nickel surfaces. II. Atomic chemisorption on Ni(100)', *Theor. Chim. Acta* **74**, 167.

[21] Mattsson, A., Panas, I., Siegbahn, P., Wahlgren, U., and Åkeby, H.: 1987, 'Model studies of the chemisorption of hydrogen and oxygen on Cu(100)', *Phys. Rev.* **36**, 7389.

[22] Bagus, P.S., Hermann, K., and Bauschlicher, C.W.: 1984, 'A new analysis of charge transfer and polarization for ligand-metal bonding: model studies of Al_4CO and Al_4NH_3', *J. Chem. Phys.* **80**, 4378.

[23] Wahlgren, U., Pettersson, L.G.M., and Siegbahn, P.: 1989, 'Cu $3d$ covalency in chemisorption?', *J. Chem. Phys.* **90**, 4613.

[24] Wahlgren, U., Pettersson, L.G.M., and Siegbahn, P.: Unpublished results.

[25] Wahlgren, U., Almlöf, J., and Siegbahn, P.E.M.: 1990, 'Oxygen chemisorption on metal surfaces using the cluster model. Basis set effects.', *Teor. Chim. Acta*, in press.

[26] Panas, I., Siegbahn, P., and Wahlgren, U.: 1987, 'Model studies of the chemisorption of hydrogen and oxygen on nickel surfaces. I. The design of a one electron effective core potential which includes $3d$ relaxation effects', *Chem. Phys.* **112**, 325.

[27] Müller, W., Flesch, J., and Meyer, W.: 1984, 'Threatment of intershell correlation effects in *ab initio* calculations by use of core polarization potentials. Method and application to alkali and alkaline earth atoms', *J. Chem. Phys.* **80**, 3297.

[28] Partridge, H., Bauschlicher, C.W., Pettersson, L.G.M., McLean, A.D., Liu, B., Yoshimine, M., and Komornicki, A.: 1990, 'On the dissociation energy of Mg_2', *J. Chem. Phys.*, in press.

[29] Pettersson, L.G.M. and Åkeby, H.: 1990, 'Core-valence correlation effects using approximate operators', *J. Chem. Phys.* **94**, 2968.

[30] Pettersson, L.G.M., Åkeby, H., Siegbahn, P., and Wahlgren, U.: 1990, 'The effects of core($3d$) correlation on chemisorption', *J. Chem. Phys.* **93**, 4954.

[31] Ahlrichs, R., Scharf, P., and Ehrhardt, C.: 1985, 'The coupled pair functional (CPF). A size-consistent modification of the CI(CD) based on an energy functional', *J. Chem. Phys.* **82**, 890.

[32] Ertl, G.: 1979, Ch. 5. in *The Nature of the Surface Chemical Bond*, T.N. Rhodin and G. Ertl (eds.), North-Holland, Amsterdam.

[33] Schüle, J., P. Siegbahn, and U. Wahlgren: 1988, 'A Theoretical Study of Methyl Chemisorption on Ni(111)', *J. Chem. Phys.* **89**, 6982.

[34] Pettersson, L.G.M. and P.S. Bagus: 1986, 'Adsorbate Ionicity and Surface Dipole Moment Changes; Cluster Model Studies of Cl/Cu(100) and F/Cu(100)', *Phys. Rev. Lett.* **56**, 500.

[35] Siegbahn, P.E.M., L.G.M. Pettersson, and U. Wahlgren: 1991, 'Theoretical Study of Atomic Fluorine Chemisorption on the Ni(100)', *J. Chem. Phys.* **94**, 4024.

[36] Egelhoff, W.F. Jr.: 1984, 'Core-level Binding-Energy-Shift Analysis of Adsorption and Dissociation', *Phys. Rev.* **B29**, 3681.

[37] Brennan, D., D.O. Hayward, and B.M.W. Trapnel: 1960, 'The Calorimetric Determination of the Heats of Adsorption of Oxygen on Evaporated Films', *Proc. Roy. Soc.* **A256**, 81.

[38] Shustorovich, E.: 1986, 1989, 'Chemisorption Phenomena: Analytical Modeling Based on Perturbation Theory and Bond-Order Conservation', *Surf. Sci. Rep.* **6**, 1; 'The Bond Order Conservation Approach to Chemisorption and Heterogenous Catalysis: Applications and Implications, *Adv. Catalys.* **37**, 1.

[39] Siegbahn, P.E.M. and U. Wahlgren: 1992, 'A Theoretical Study of Atomic Oxygen Chemisorption on the Ni(100) and Ni(111) Surfaces', *Intern. J. Quantum Chem.*, in press.

ELASTIC SCATTERING OF ATOMS FROM SOLID SURFACES. THE ^{4}He—Cu(11α) $(\alpha = 0, 3, 5, 7)$ EXAMPLE*

S. MIRET-ARTÉS, M. HERNÁNDEZ**, J. CAMPOS-MARTÍNEZ,
P. VILLARREAL, and G. DELGADO-BARRIO
Instituto de Física Fundamental
Centro Mixto C.S.I.C.-U.C.M.
Serrano 123
28006 Madrid
Spain

ABSTRACT. A very exhaustive analysis of selective adsorption resonances, for the ^{4}He—Cu(11α) systems, is presented. Emphasis on results, supplied by the Close-Coupling and Golden-Rule formulations, is made with different types of elastic scattering: non-resonant, resonant, critical, at threshold and temperature dependent conditions. Connected with the critical kinematic effect, new results are predicted.

I. Introduction

Nowadays, it is out of the question the importance of gas-surface interactions. Progress in technological problems and development of theoretical and computational methods have motivated this increasing interest. However, this attractive situation did not take place at the very beginning of this field. Mainly, two factors contributed to make difficult studies about these interactions. Firstly, the nonreproducibility of many experimental results due to the difficulty of getting identical structures on solid surfaces; and second, the way of how experiments were carried out. The solid surface was totally inmersed in the gas in thermal equilibrium with it. Thus, only average properties were possible to be measured. But, in spite of these drawbacks, important conclusions were obtained. Studies about the "accomodation coefficient" (defined, firstly by Knudsen [1], as a measure of the efficiency of the energy exchange at an interface between a gas and a solid at different temperatures) and the statement of the cosine distribution as a universal law in gas-surface scattering were two chief achievements. Nevertheless, in 1928, evidence for the specular intensity of molecular beams was established. Although the molecular beam was invented about 1911, its application to the study of gas-surface interactions began with Stern and his group, providing the first proof of the wave nature of atomic and molecular beams [2]. Also, the selective adsorption phenomenon was discovered by Stern, leading to the first quantum

* Work supported by CICYT under Grant No. PB87-0272 and EEC Grant No. SC1.145.C
** Also: Dpto. de Física Fundamental y Experimental. Univ. de La Laguna, 38203 La Laguna, Sta. Cruz de Tenerife, Spain.

L. A. Montero and Y. G. Smeyers (eds.), Trends in Applied Theoretical Chemistry, 19–50, 1992.

mechanics treatment of this resonance phenomenon by Lennard-Jones and Devonshire [3]. It occurs when the z-component of the kinetic energy of the atomic/molecular stream becomes equal to the energy of a bound state of the normal motion potential function. These authors already pointed out the importance of diffraction in yielding information on gas-surface interaction potentials. After this very fruitfull period, very few works were published during at least two decades. Those works were mainly devoted to experimental studies of accomodation coefficients.

During the 1960s, improvements about vacuum conditions and methods of monitoring surfaces were widely developed and, therefore, a new interest in this research field grew rapidly. In these molecular beam scattering experiments, high or ultrahigh vacuum conditions are necessary in order to maintain the surface cleaned, since background gas is seldom the only source of surface contamination, apart from adsorption of gas from the molecular beam and manufacturing process and history of the target sample. Thus, low-energy electron diffraction (LEED) techniques [4] and Auger electron spectroscopy (AES) [5] turned out to be very useful tools for controlling solid geometries and identifying surface atoms, respectively. Actually, they are very often incorporated into the molecular beam apparatus [6]. With this way open, theoretical works increased quickly. The earliest ones in modern classical theory began with Cabrera, Swanzig and Goodman [7], three-dimensional lattice model Monte-Carlo trajectory calculations were made by Oman and McClure [8], and Logan and Stickney proposed the first of the classical "cubes" models [9]. At the same time, modern quantum theory began with Tsuchida [10], Cabrera, Celli, Goodman and Manson (CCGM) [11], and Wolken [12]. Models made of the gas, of the solid and of the interaction potential between the two partners, and used in the theory, have not been highly improved in the last two decades.

The great avalanche of experimental results gathered as well as the experience due to the first classical-mechanical trajectory calculations of gas-surface scattering led to set up a classification of this scattering. Experimental results are often described in terms of scattering patterns. Thus, if a molecular beam is directed to the surface with (θ_i, ϕ_i)-angles of incidence (polar and azimuthal angles of the incident beam, respectively), the scattered beam intensity can be collected with a detector at (θ_f, ϕ_f)-angles. Usually, ϕ_f is taken to be zero and only (θ_i, θ_f)-angles are measured (incidence plane). According to the shape of these diagrams or scattering patterns, these processes can be brought under different types: diffuse, lobular, rainbow, etc. However, it is recognized a more general classification of experimental results based on three physical regimes [1, 8]: thermal, structure and penetration and sputtering. At incident energies, E_i, above 40 eV, predominates the penetration and sputtering regime. At thermal energies, the thermal motions of the surface atoms dominate the scattering process whereas, in the structure regime, the roughness or corrugation of the surface does it. In both these regimes, elastic and inelastic processes can occur. In order to differentiate them, a set of dimensionless parameters are defined [1]. The most important ones are the following: the mass ratio ($\mu = M_g/M_s$); the radius ratio ($\mathcal{R} = R/R_c$), R being the distance of closest approach of the center of a gas atom to that of a surface atom during a collision, and R_c a critical value; and the energy ratio ($\mathcal{E} = E_i/k_B\Theta_C$), with k_B the Boltzman constant and Θ_C being a characteristic vibration temperature of the solid and it is related to the maximum modal frequency, $\Omega_{\max}$, of the solid by $\hbar\Omega_{\max} = k_B\Theta_C$. Clearly, μ gives us an indication of whether single or multiple collisions of gas atoms with surface are important, $\mathcal{R}$ is a measure of the surface roughness and $\mathcal{E}$ is a criterion for determining if a given scattering is inelastic or not. The energies exchanged will be of order $k_B T_g$ and interaction times of order $a(M_g/k_B T_g)^{1/2}$ with a a typical interatomic spacing. At thermal energies (10 meV–5 eV), we can speak about two processes [13]: trapping and sticking. This difference is based on the nature of the attractive interaction. In the first case, the physical adsorption or physisorption (by van der Waals forces) takes place when binding energies are between 5 and 50 KJ/mol. In the second case, the chemisorption, binding energies are between 50 and 500 KJ/mol. Although this boundary between both processes is somewhat arbitrary, it is useful from a conceptual point of view. The sticking process can occur directly through an activation energy or indirectly via a trapping mechanism. The particles adsorbed at the surface are called ad-particles (adatoms, etc.). Residence times, desorption phenomena via phonon and/or photon interaction are among the most important topics in this field. Furthermore, we have the elastic scattering with no energy or parallel momentum exchange (specular scattering), with parallel momentum exchange (diffraction) and with energy exchange with internal states of the gas particle.

Obviously, more situations can be devised at surfaces but, in this paper, we are concerned with elastic scattering of light atoms by metalic solid surfaces. Transfer of thermal incident energy to kinetic energy parallel to the surface, with parallel momentum exchange, can lead to a "virtual" trapping [13b] as in the selective adsorption phenomenon (where the gas particle will simply bounce off) and contrary to a "real" trapping where "stay-times" are similar to typical experiment times. If these stay-times or residence times are non-negligible, the "virtual" trapping could add to the probability of energy exchange with the surface and could be considered as a precursor step for trapping and/or sticking. Moreover, recently, we have pointed out [14] manifestation of selective adsorption resonances (SAR), in diffraction spectra, is very dependent on scattering geometry, allowing to control experimentally optimum appearances. Besides, at certain experimental parameter values (incident energy, E, polar, θ, and azimuthal, ϕ, angles of the incident wave vector of the particle), called by us "*critical conditions*", it is possible to enhance resonance features in such a way that stay-times are minimized. This predicted critical kinematic effect is very similar to the kinematic focussing effect [15], appearing in angular spectra of inelastic scattering.

Before concluding this Section, we would like to mention the standard notations and conventions. These are those of [11]. The surface is parallel to the xy-plane. The z-direction is chosen as the outward normal to the surface. Vectors, in three-dimensional space, are denoted by lower case letters while capital letters are used for vectors parallel to the surface. Thus,

$$
\begin{aligned}
\underline{r} &= (x, y, z) & &= (\underline{R}, z) \\
\underline{k} &= (k_x, k_y, k_z) & &= (\underline{K}, k_z) \\
& & &= (k \sin\theta \cos\phi,\ k \sin\theta \sin\phi,\ k \cos\theta) \\
k &= |\underline{k}| = \tfrac{2M_g}{\hbar^2} E
\end{aligned}
$$

with M_g the mass of the gas-particle.

Although any discussion about gas-surface interactions should include simultaneously knowledge from experiment, interaction potentials and dynamical methods, conceptually, this is not a very convenient way of presentation. Thus, the paper is organized as follows. In Section II, an outline of molecular beams and detection techniques is given. In Section III, the most widely used model potentials for elastic scattering are described. In Section IV, a general account of the two theoretical formalisms (energy and time dependent) is presented, with a special emphasis to the Close-Coupling formulation. Finally, in Section V, new results for the ^{4}He—Cu(11α) systems (with $\alpha = 0, 3, 5, 7,$) are presented and discussed. At the end of the three first Sections, references to these systems are supplied.

II. Brief Remarks on Experimental Techniques

In this Section, we would like to comment some of the experimental problems in molecular beams and detection techniques. Detailed information can be found in [1], [13b], and [16].

Two types of molecular beams are used in scattering experiments: effusive and nozzle source beams. In effusive source beams, the source is enclosed in a larger chamber, which is steadily evacuated by pumps, and presents a small aperture in one of its walls. The effusive stream exhibits a free-molecular behavior in such a way that the Knudsen number, K_n, is much greater than one. K_n is defined as the ratio of mean free path to the smallest dimension of the source aperture. The molecular beam is formed from an effusive stream by collimation. The velocity distribution for the flux on the beam axis is given by

$$g(c) = \frac{1}{2}(M_g/k_BT_g)^2 c^3 \exp(-M_g c^2/2k_BT_g) \tag{II.1}$$

and $(9\pi k_BT_g/8M_g)^{1/2}$ and $2k_BT_g$ are, respectively, the mean speed and energy. Maximum mean molecular speed attained with these sources is typically of order 10^5 cm/s. The spatial distribution is governed by the cosine law ($cos\theta d\Omega$, with Ω denoting the initial direction (θ_i, ϕ_i)). The temperature of the beam is T_g due to the free-molecular behavior of the effusive stream. The source and target chambers are pumped separately with typical pressures of 10^{-5} and 10^{-7} Torr, respectively, in order to avoid attenuation of the beam by collisions with background gas. However, although the properties of the beam may be calculated quite precisely, its broad speed distribution and divergences on the direction of molecular motion are the major disadvantages of the effusive beam.

Thus, due to these drawbacks, nozzle source beams were used mainly in experiments. Narrower velocity distributions and greater intensities than in effusive beams are obtained. They are generated by skimming the core of high Mach number, low-density jets and where a nozzle on the source orifice, a conical skimmer on the first collimating slit and a third pumped chamber for the target are the new components. A supersonic free jet is allowed to expand through a sonic nozzle from a source at high pressure (say, 10 atm) into an evacuated chamber at $\sim 10^{-4}$ Torr and to a second one, with 10^{-6} Torr, through the conical skimmer. Only the axial core of the jet enters into this second chamber. The target chamber, also at 10^{-6} Torr, receives the beam. Very high Mach number is not always desirable in rapidly expanding gas because partial condensation is possible. Because of cooling during the expansion, T is much smaller than T_g. If a gas mixture is used (seeded beam) to form a supersonic jet, mean molecular speed is similar to a uniform gas expanded from a source at a given temperature. For helium atoms, kinetic energies in the range 20 to 300 meV are easily achieved. Measurements of the properties of this beam are not an easy task since transition from continuum (high Reynolds number or very small Knudsen number) to free-molecular behavior of the stream is presented. The velocity distribution for a supersonic beam can be written as [13b]

$$g(c) \propto c^3 \exp\left(-\frac{M_g}{2k_BT}(c - c_0)^2\right) \tag{II.2}$$

where c_0 is a non-zero flow velocity and T is the characteristic temperature of the spread of parallel velocities. Beam divergences of $0.3°$ are considered satisfactory. However, standards for generating beams and calibrating vary from laboratory to laboratory. As a consequence [1], quantitative comparisons among sets of data cannot always be carried to the same degree of detail. More important is to compare trends in experimental data.

Detection techniques [13b] in molecular beam scattering experiments are, very often, mass spectrometers with electron bombardment ionizers and bolometers. If internal state detection is required, laser induced fluorescence, multiphoton ionization and infrared excitation can be used. Doppler and time-of-flight measurements can also be made to obtain velocity information. Residence times can be measured if the beam is chopped with a high-speed rotating shutter, controlling the time delay from opening the shutter to arrival of the beam at the detector. In general, detector as well as sample are made movable and source is fixed.

The experimental set-up, used by Lapujoulade *et al.*, for studying experimentally the scattering of ^{4}He atoms from different copper surface faces, $Cu(11\alpha)$ with $\alpha = 0, 3, 5, 7$, has been fully described in [17a]. A Campargue nozzle beam with the following characteristics is obtained: mean velocity of 1.76×10^5 cm/s, mean energies ranging from 20 to 65 meV, Mach number of 16, intensity molecular flow of 10^{20} atoms/steradian, velocity distribution $\Delta c/c = 0.06$ and divergency of 0.1×10^{-3} steradian. The sample has a diameter of 20 mm and a thickness of 1 mm. θ_i may be adjusted from $10°$ to $90°$ and surrounded by a 5×10^{-11} Torr background pressure. The sample temperature, ranging from ambient to melting point, is regulated by an electron gun located on its back side. An Auger spectrometer is introduced in order to check the cleanliness of the surface. The detection apparatus has been an electron impact ionization 380 nm away from the sample.

Typically, scattered intensity (for different diffraction peaks) is recorded as a function of the reflection angle, θ_r; θ_i and beam energy being fixed. Also, temperature dependence of these scattered intensities between $70\ K$ (inelastic scattering is not very probable) and $770\ K$ was examined, following a linear behavior at low temperatures. Usually, these results were extrapolated to zero temperature in order to compare with elastic calculations together with different model potentials. The unitarity of the intensity recorded is taken to be the sum of all peak intensities measured in the incidence plane (in-plane unitary). This unitary becomes worst as copper surface faces are more open ($\alpha = 5$ and 7). Random surface defects and out-of-plane scattering (due to out-of-plane corrugation) are mainly responsible for this decreasing. However, in-plane intensities were normalized to unity and a one-dimensional corrugation model was proposed. Resonance signatures and widths (angular and energetic ones) have also been reported for some of those surfaces. [17] gathers all this information.

III. Gas-Surface Interaction Model Potentials

One of the main goals of these scattering experiments is to extract information about the interaction between a gas particle and a solid surface. Clearly, depending on the scattering regime studied, the gas-surface interaction will present different interesting aspects. For example, at thermal energies, monoatomic molecular beams can be utilized as probes for surface properties because gas particles do not penetrate beyond the first surface layer of a solid. If inelastic scattering is predominant, a model for the motion of solid nuclei is required. In any case, interaction among gas particles of the stream is assumed to be negligible. Only position and orientation of those particles are considered. In general, gas-surface interaction model potentials are obtained by extrapolation of experimental data to zero surface temperature. Temperature dependence on diffracted intensities is taken into account by means of the Debye-Waller factor [11], which gives us the ratio between scattering intensity at temperature T and the corresponding one at zero temperature.

The potential energy function must contain information about the position and orientation of the gas particle, $\underline{r}$, and the set of displacements of all the solid atoms, $\underline{u}$. Thus, we can denote [11] this function as $V(\underline{r}, \underline{u})$. If this function is averaged over the thermal motion of the solid atoms, we have

$$V(\underline{r}) = \langle V(\underline{r}, \underline{u}) \rangle \tag{III.1}$$

This average $V(\underline{r})$ is the function taken for studying elastic scattering. Consideration of $V(\underline{r}, \underline{u})$ or inelastic scattering will not be treated here. The reader interested in this subject can consult [1], [11], [13], [16], [18], and [19].

Different forms can be proposed for $V(\underline{r})$ [20]. In general, this function can be represented by a suitable summation of pairwise potentials between the gas particle position, $\underline{r}$, and the positions of solid atoms, $\underline{r}_i$,

$$V(\underline{r}) = \sum_i V_i \left(|\underline{r} - \underline{r}_i| \right) \tag{III.2}$$

This summation can sometimes lead to analytical results or can be replaced by an integral when a continuum model of the solid is assumed (flat surface).

These pairwise potentials can conceptually be separated into three regions: attractive, intermediate and repulsive. As it is well known, the short-range or repulsion region is dominated by the overlapping between electronic wave functions of the gas particle and those corresponding to the solid surface. The long-range or attractive region is governed by the van der Waals (vdW) interaction considering, in this case, the crystal as a continuum with a certain dielectric function. At intermediate range, near the potential minimum, this interaction is represented by a function capable of fitting the bound or adsorption states. A complete discussion of the analytical forms more convenient in each region can be found in [13a], [20], and [21]. However, the so-called ESMSV potential (exponential-spline-Morse-spline-van der Waals) developed from crossed-beam scattering data has turned out to be very adequate. It can be expressed as

$$V \left(|\underline{r} - \underline{r}_i| \right) \equiv V(\rho) = \tag{III.3}$$

$$= \begin{cases} A \exp[-2\chi(\rho - \rho_i^e)], & \rho \ll \rho_i^e \\ D\{\exp[-2\chi(\rho - \rho_i^e)] - 2\exp[-\chi(\rho - \rho_i^e)]\}, & \rho \sim \rho_i^e \\ -\frac{C_6}{\rho^6} - \frac{C_8}{\rho^8} \cdots \cdots, & \rho \gg \rho_i^e \end{cases}$$

with ρ_i^e the equilibrium position and $A, D, \chi, C_6, C_8 \cdots$ parameters to be fitted.

For a perfect periodic solid surface, $V(\underline{r})$ can be expanded in Fourier series as

$$V(\underline{r}) = \sum_{\underline{G}} V_{\underline{G}}(z) \exp[i\underline{G} \cdot \underline{R}] \tag{III.4}$$

$\underline{G}$ being the two-dimensional reciprocal lattice vector. For $\underline{G} = \underline{0}$, $V_{\underline{0}}(z)$ is the so-called laterally averaged potential. The eigenvalues of $V_{\underline{0}}(z)$ are the surface adsorbed states. Again, typical model potentials can be assumed for this first coefficient. The $V_{\underline{G}}(z)(\underline{G} \pm \underline{0})$ are the "diffracting potentials" or couplings giving rise to diffraction phenomenon. These coefficients are calculated according to the following formula

$$V_{\underline{G}}(z) = S^{-1} \int_S V(\underline{r}) \exp[-i\,\underline{G} \cdot \underline{R}]\, d^2\underline{R} \tag{III.5}$$

where S is the area of the unit cell. The effective range of these coefficients is short for all $V_{\underline{G}}(z)(\underline{G} \pm \underline{0})$, depending on the diffraction order.

Ab initio calculations have also been performed for this interaction. However, application of the standard methods of electronic calculation has been limited in view of the difficulty presented by the contribution of the electronic clouds of the solid. In [13b] and [22] can be found more information about these calculations. Only we would like to stand out the work of Esbjerg and Nørskov [23] about the repulsion part of the $V(\underline{r})$ potential. These authors suggested that this part could be taken as proportional to a certain average of the electronic charge density due to the surface alone. This idea was applied by Hamann [24] for calculating the corrugation associated with the electron density at a metal surface. When a stream of gas atoms is directed to the surface with high incident energies, the penetration of these atoms is deeper and, therefore, they feel a higher electron density and, consequently, the effective corrugation will increase. Although the agreement between theoretical and experimental results was not very good, it was the first time that a correspondence between this electron density and the parameters governing geometrically the corrugation of the surface was established.

For molecules, *ab initio* calculations are even less frequent. Only for a diatomic, the potential function depends on five or six variables whether the bond distance is considered fixed or not, respectively. A high number of parameters is required in order to reduce the potential to a three-dimensional function. LEPS (London-Eyring-Polanyi-Sato) semiempirical potentials have been widely used in this context. The corrugation of the surface is very often ignored or considered in a very simple way [12b, 25].

Garibaldi *et al.* [26] proposed a very simple interaction model, leading to very good results when the corrugation is small. The hard corrugated wall potential is expressed as

$$V(\underline{r}) = \begin{cases} 0, & z > \varphi(\underline{R}) \\ \infty, & z \leq \varphi(\underline{R}) \end{cases} \tag{III.6}$$

where the surface is represented by the shape function $\varphi(\underline{R})$. If the surface is periodic, again, $\varphi(\underline{R})$ can be expanded by a Fourier series

$$z = \varphi(\underline{R}) = \sum_{\underline{G}} \varphi_{\underline{G}} \exp[i\underline{G} \cdot \underline{R}] \tag{III.7}$$

The simplest form of the shape function is

$$\varphi(\underline{R}) = h \left[\cos \left(\frac{2\pi x}{a} \right) + \cos \left(\frac{2\pi y}{a} \right) \right] \tag{III.8}$$

where $\varphi_{1,0} = \varphi_{0,1} = \varphi_{-1,0} = \varphi_{0,-1} = \frac{1}{2}h$ and the corrugation period is a. All $\varphi_{\underline{G}}$ are called corrugation parameters. The values of these parameters are evaluated from the experimental diffracted intensities. To do this, several dynamical methods are used. In the next Section, some of them will be presented. A detailed exposition of potential and corrugation parameters can be found in [13a].

For the system we are interested here, several model potentials have been fitted. Apart from the hard corrugated wall, we have the hard corrugated wall with a well [27], the corrugated exponential potential [28],

$$V(\underline{r}) = \exp\left\{-2\chi\left[z - \phi(\underline{R})\right]\right\} \tag{III.9}$$

the corrugated Morse potential (CMP) [29],

$$V(\underline{r}) = D\left\{V_{\underline{0}}^{-1}\exp\left[-2\chi\left(z - \phi(\underline{R})\right)\right] - 2\exp\left[-\chi z\right]\right\} \tag{III.10}$$

and the modified CMP (MCMP) [17f]. The difference with the previous one comes out with the diffracting potentials.

In Eq. (III.10), D, χ and $V_{\underline{0}}$ are the well depth, the stiffness parameter of the Morse potential and the surface average of $\exp[2\chi\varphi(\underline{R})]$ on the unit cell. If Eq. (III.10) is expanded in Fourier series, the different coefficients of this series are expressed as

$$V_{\underline{0}}(z) = D\left\{\exp\left[-2\chi z\right] - 2\exp\left[-\chi z\right]\right\}$$

$$V_{\underline{G}}(z) = D\frac{V_{\underline{G}}}{V_{\underline{0}}}\exp\left[-2\chi z\right] \quad, \quad \underline{G} \pm \underline{0} \tag{III.11}$$

with

$$V_{\underline{G}} = S^{-1}\int_S \exp\left[2\chi\varphi(\underline{R})\right]\exp\left[-i\underline{G}\cdot\underline{R}\right]d^2\underline{R} \tag{III.12}$$

The CMP cannot be applied to the whole experimental energy range with the same set of parameters. For each incident energy, $\varphi(\underline{R})$ must be modified since the effective corrugation of the surface changes appreciably. This was the reason for which the MCMP was proposed.

In [17f], it can be found a detailed discussion of the effective corrugation, the CMP and MCMP and the potential parameters convenient for the ^{4}He—Cu(11α) systems. For our purposes, we have taken the following values for the CMP: D = 6.35 meV, $\chi = 1.05$ Å^{-1}. The coefficients $V_{\underline{G}}$ are those given in Table 9 of [17f]. As it was already said, the different copper faces display, in a first approximation, a corrugation to be only significant in the x-direction. Thus, for these coefficients, $\underline{G} = \frac{2\pi}{a}(m, 0)$, a being the unit cell length. This value changes with the copper surface face: $a = 3.6$ Å($\alpha = 0$), $a = 4.227$ Å($\alpha = 3$), $a = 6.625$ Å($\alpha = 5$), and $a = 9.12$ Å($\alpha = 7$). Up to four values of $V_{\underline{G}}$ are necessary in the calculations for the more corrugated faces ($\alpha = 5$ and 7). For $\alpha = 0$ and 3, only two real coefficients are important.

If the temperature of the surface is taken into account in the full dynamical problem, CCGM [11] devised a way of obtaining thermally-averaged potential energy functions from the instantaneous potential functions based on the DW theory. It can be shown that Eq. (III.11) is modified when a thermal average is carried out. Thus, we have

$$V_{\underline{0}}^T(z) = D^T\left\{\exp\left[-2\chi(z - z_0^T)\right] - 2\exp\left[-\chi(z - z_0^T)\right]\right\}$$

$$V_{\underline{G}}^T(z) = A_{\underline{G}}^T D^T \exp\left[-2\chi(z - z_0^T)\right] \tag{III.13}$$

with

$$D^T = D \exp\left(-\chi^2 \langle u_z^2 \rangle_T\right)$$

$$z_0^T = \frac{3}{2}\chi \langle u_z^2 \rangle_T \tag{III.14}$$

$$A_{\underline{G}}^T = \frac{V_{\underline{G}}}{V_{\underline{0}}} \exp\left(-\frac{1}{2}\underline{G}^2 \langle u_z^2 \rangle_T\right)$$

$\langle u_z^2 \rangle_T$ being a mean square lattice displacement, averaged on temperature. It has been assumed here that the displacement in x and z directions are the same. Evaluation of these displacements of lattice atoms is not an easy task and several corrections have been applied for this goal. Usually, a Debye continuum model for the thermal vibrations of the solid is supposed. In our case, we have taken the values given in [17d], which were determined assuming a function, more or less complicated, of the temperature with adjustable parameters in order to make a good fit to the experimental data. Eqs. (III.13) and (III.14) have been only used for the resonant scattering of the ^{4}He—Cu(110) system.

IV. Theoretical Methods

Diffraction experiments, in resonance conditions or not, provide an adequate procedure in order to propose model interaction potentials as well as to obtain valuable information about the structure of surfaces. These goals are achieved by means of any dynamical method. Most of theoretical approximations or methods used come from atomic and nuclear physics and have been widely developed in the field of molecular dynamics of weakly bound systems. Broadly speaking, two theoretical formalisms have been applied to the elastic scattering treatment: time independent and time dependent formalisms. In the first one, some of methods developed stem from classical theory of diffraction, mostly in optics and surface science. In our opininion, it is not risky to say that one of the main purposes of the great variety of theoretical treatments, apart from its intrinsic interest, is to overcome the severe computing time limitation of the Close-Coupling calculations. However, the Close-Coupling formulation remains nowadays as the "exact" treatment in time independent formulations.

We proceed now to discuss briefly the main theoretical methods.

IV.1. TIME INDEPENDENT FORMALISM

IV.1.1. The Close-Coupling (CC) Formulation. Tsuchida [10] and Wolken [12] were the firsts in carrying out CC calculations in gas-surface scattering. Subsequent CC calculations were carried out by Chow and Thompson [30a]. Chow also developed an iterative integration scheme using the Green function for solving the coupled equations [30b].

Consider the elastic scattering of a gas atom from a statistically corrugated periodic solid surface. If the atom is an structureless particle of mass M_g and incident wave vector $\underline{k}$, the corresponding Schrödinger equation can be written as

$$\left[-\frac{\hbar^2}{2M_g}\nabla_{\underline{r}}^2 + V(\underline{r}) - \frac{\hbar^2}{2M_g}\underline{k}^2\right]\Psi(\underline{r}) = 0 \tag{IV.1}$$

Due to the periodicity of the surface lattice, $V(\underline{r})$ and $\Psi(\underline{r})$ can be expanded in Fourier series. Similarly to Eq. (III.4), we have

$$\Psi(\underline{r}) = \sum_{\underline{G}} \psi_{\underline{G}}(z) \exp[i(\underline{K} + \underline{G}) \cdot \underline{R}]$$

(IV.2)

Substituting Eqs. (III.4) and (IV.2) into (IV.1), and after a little algebra, we obtain the usual set of coupled equations for the diffracted waves

$$\left[\frac{\hbar^2}{2M_g} \frac{d^2}{dz^2} + \epsilon_{\underline{G}}(E, \theta, \phi) - V_{\underline{0}}(z) \right] \psi_{\underline{G}}(z) = \sum_{\underline{G}' \neq \underline{G}} V_{\underline{G} - \underline{G}'}(z) \psi_{\underline{G}'}(z)$$

(IV.3)

$\epsilon_{\underline{G}}$ being the z-component of the $\underline{G}$-diffraction channel kinetic energy, expressed as a function of the incident or geometric parameters (E, θ, ϕ) of the scattering as

$$\epsilon_{\underline{G}}(E, \theta, \phi) = E - (\hbar^2/2M_g) \left| (2M_g E/\hbar^2)^{1/2} \sin\theta(\cos\phi, \sin\phi) + \underline{G} \right|^2$$

(IV.4)

with $\underline{K} = (2M_g E/\hbar^2)^{1/2} \sin\theta(\cos\phi, \sin\phi)$. Each $\underline{G}$-channel is represented by an effective potential of the form $\left[(2M_g/\hbar^2)V_{\underline{0}}(z) + (\underline{K} + \underline{G})^2 \right]$ and where $(\hbar^2/2M_g)(\underline{K} + \underline{G})^2$ can be seen as the asymptotic energy of the $\underline{G}$-channel. The eigenvalues of $V_{\underline{0}}(z)$ are the surface adsorbed states or selective adsorption resonance positions at zero order. $V_{\underline{G} - \underline{G}'}(z)$ are responsible for the shifts and widths of those resonances. The number of $\underline{G}$-channels involved in the calculations depends strongly on the surface corrugation or strength of Fourier coefficients. Time required in solving Eq. (IV.3) is proportional to N^3, N being the number of $\underline{G}$-channels (closed, $\epsilon_{\underline{G}} < 0$, or open, $\epsilon_{\underline{G}} > 0$). Eq. (IV.3) is solved subject to the usual boundary conditions [12, 31] defining the scattering S-matrix. Observable diffraction intensities are proportional to $| S_{\underline{G}0} |^2$, starting from the specular channel $(\underline{G} = \underline{0})$.

A selective adsorption resonance (SAR) fulfils the following condition,

$$\epsilon_{\underline{\rho}}(\bar{E}, \bar{\theta}, \bar{\phi}) = \varepsilon_{\nu} < 0$$

(IV.5)

where $\underline{\rho} \in \underline{G}$ and denotes the resonant diffraction channel (closed) and ε_{ν} is the ν eigenvalue of $V_{\underline{0}}(z)$. This SAR is usually denoted by $\binom{m}{\nu}{n}$; m, n being the components of the $\underline{\rho}$-vector in a given basis of the reciprocal lattice. For each SAR, a mathematical surface of different resonant scattering geometries $(\bar{E}, \bar{\theta}, \bar{\phi})$ can be calculated according to Eq. (IV.5). In these conditions, diffraction intensities display a sharp peak around the resonant position. The corresponding S-matrix can be expressed as [32, 33]

$$S\left(\{ \epsilon_{\underline{G}}(E, \theta, \phi) \}_{\underline{G}} \right) =$$

$$S_{bg}\left(\{ \epsilon_{\underline{G}}(E, \theta, \phi) \}_{\underline{G}} \right) \left[I - \frac{i\,\mathcal{A}}{\epsilon_{\underline{\rho}}(E, \theta, \phi) - \bar{\varepsilon}_{\nu} + i\Gamma_{\rho\nu}/2} \right]$$

(IV.6)

S_{bg} is the background S-matrix, $\bar{\varepsilon}_{\nu}$ and $\Gamma_{\rho\nu}$ are the position and width of the SAR, respectively. $\mathcal{A}$ is a complex matrix whose trace is $\Gamma_{\rho\nu}$, and I the unit matrix. With

$\{\cdots\}_{\underline{G}}$ we are meaning that the independent variables of the S and S_{bg} matrices are the kinetic energies of all $\underline{G}$-channels. In Eq. (IV.6), it has been supposed that the resonance is isolated, simple and non-degenerate on ϵ_p, which is considered the resonant dynamical variable. Obviously, the arrangement of these diffraction channels is quite different [31, 33] depending on the scattering geometry. Thus, we can speak about a "moving threshold" multichannel scattering, in contrast to other types of scattering (e.g., in atomic or molecular physics) where thresholds are fixed.

In a previous paper [33], we have shown SAR positions and widths are very dependent on the experimental parameters (E, θ, ϕ). $\Gamma_{\rho\nu}$ can not be directly measured since, experimentally, it is only possible to vary these parameters and not ϵ_p. We can say that these widths are "inherent" widths, associated to internal variables of the theory. We proposed an explicit relation between these widths and those associated with the experimental parameters, which can be directly observed in a diffraction spectrum. Thus, we have

$$\Gamma_{\underline{\rho}\nu}/2 = | \ \epsilon_{\underline{\rho}}(\bar{E} \pm \Gamma_E^{\pm}/2, \bar{\theta}, \bar{\phi}) - \varepsilon_\nu \ | \simeq \left| \left(\frac{\partial \epsilon_{\underline{\rho}}}{\partial E} \right)_{\bar{E}, \bar{\theta}, \bar{\phi}} \cdot \Gamma_E^{\pm}/2 \right| \qquad \text{(IV.7)}$$

Similar relations can be written for the angular half-widths, $\Gamma_\theta^{\pm}/2$ and $\Gamma_\phi^{\pm}/2$ ($\pm$ signs are used for defining right-hand and left-hand variations in the corresponding variable; in general, widths are asymmetric). In these widths, the subindex ν is understood. Moreover, at certain resonant incidence conditions

$$\left(\left(\frac{\partial \epsilon_{\underline{\rho}}}{\partial E} \right)_{\bar{E}, \bar{\theta}, \bar{\phi}} = 0, \text{"critical conditions"} \right)$$

a very important enhancement of Γ_E occurs. Analogously, this situation can also take place for the ϕ angle. This effect is the counterpart of the kinematic focussing effect, reported by G. Benedek [34], appearing in angular spectra of inelastic scattering. As a consequence of this new analysis, optimum appearances of SAR in diffraction spectra can be controlled experimentally by means of a detailed study of the elastic scattering kinematic conditions.

IV.1.2. Sudden Approximations.   Gerber *et al.* [35] proposed an adaptation of the sudden approximation (familiar in gas-phase scattering) to atom-surface scattering. This approach requires little computational effort as compared with CC calculations. Essentially, this approximation is applied when $k_z \gg | \ \underline{G} \ |$, for all diffraction states significantly populated in the scattering process. With this, an important decoupling of the multichannel equations is obtained. Several versions of this approximation can be found in [36].

IV.1.3. Golden Rule Framework.   Within this framework, several decoupling schemes have been succesfully applied to study vibrational and rotational predissociation of vdW molecules [37]. Some of these, based on a diabatic distorted wave approximation, have been used in gas-surface interactions [31].

Starting from the set of CC equations, Eq. (IV.3), a zero-order approximation for calculating positions and widths of resonances could consist of neglecting all the coupling elements, $V_{\underline{G}-\underline{G}'}(z)$. Thus, the resulting uncoupled equations are

$$\left[\frac{\hbar^2}{2M_g}\frac{d^2}{dz^2} + \epsilon_{\underline{G}}(E, \theta, \phi) - V_{\underline{0}}(z)\right]\psi_{\underline{G}}^{(0)}(z) = 0 \tag{IV.8}$$

This equation has discrete as well as continuum solutions for each $\underline{G}$. The effective potentials are formed by $V_{\underline{0}}(z)$ and the asymptotic energy corresponding to a $\underline{G}$ given. In this way, ε_ν together with this asymptotic energies give zero-order energy positions of resonances. Continuum energies, ε, are given by

$$\varepsilon = \varepsilon_\nu + \left(\frac{\hbar^2}{2M_g}\right)[(\underline{K}+\underline{\rho})^2 - (\underline{K}+\underline{G})^2] > 0$$

and represent the relative kinetic energies between two diffraction channels, $\underline{\rho}$ and $\underline{G}$. Both energies are measured from the asymptote of each $\underline{G}$-channel. The corresponding wave functions are $\psi_{\underline{\rho}\nu}^{(0)}(z) = \psi_{\underline{0}\nu}^{(0)}(z)$ and $\psi_{\underline{G}\varepsilon}^{(0)}(z) = \psi_{\underline{0}\varepsilon}^{(0)}(z)$.

Now, according to the Golden-Rule formulation, the width associated to a ν−quasibound state, sustained by the $\underline{\rho}$-channel, is estimated by means of

$$\Gamma_{\underline{\rho}\nu} = 2\pi \sum_{\underline{G}} \left|\langle\psi_{\underline{G}\varepsilon}^{(0)}(z) \mid V_{\underline{\rho}-\underline{G}}(z) \mid \psi_{\underline{\rho}\nu}^{(0)}(z)\rangle\right|^2 \tag{IV.9}$$

and where each contribution must be calculated "on-shell energy". Whenever the corrugation of the surface lattice is small, this diabatic decoupling scheme will work well. Nevertheless, this approach will be no longer adequate in those surfaces presenting high corrugation parameters since the couplings become more important. Even more, if the spacing among strongly coupled diffraction channels is less than the well depth of the $V_{\underline{0}}(z)$ potential, then discrete-discrete and continuum-continuum couplings can not be neglected in higher orders of approximation. In general, the first kind of couplings is much more relevant than the second ones. A straightforward improvement of this diabatic scheme could take into account only discrete-discrete couplings through a configuration interaction (CI) to describe more properly the adsorption states, according to the following formula

$$\Gamma_{\underline{\rho}\nu}^{CI} = 2\pi \sum_{\underline{G}} \left|\langle\psi_{\underline{G}\varepsilon}^{(0)}(z) \mid V_{\underline{\rho}-\underline{G}}(z) \mid \psi_{\underline{\rho}\nu}^{(1)}(z)\rangle\right|^2 \tag{IV.10}$$

with $\psi_{\underline{\rho}\nu}^{(1)}(z)$ the first-order wave function issued from a CI calculation among all $\underline{G}, \nu$−states.

Inclusion of continuum-continuum couplings are now in progress in our laboratory.

IV.1.4. The CCGM Formulation. This formulation is based on the two-potential T-matrix scattering theory due to Gell-Mann and Goldberger [38]. It is well known the transition rate from some initial state i to same final state f is given exactly by

$$P_{fi} = \frac{2\pi}{\hbar} \mid T_{fi} \mid^2 \delta(E_i - E_f) \tag{IV.11}$$

where E_i and E_f are the total energies of i and f states, respectively. Now, the problem is reduced to the calculation of this matrix element. Usually, this is done in terms of a reduced T-matrix, t. The final result is given by a complicated matrix integral equation. This equation has never been solved exactly but by a process of high order distorted wave perturbation [28]. The first order distorted-wave Born approximation renders the theory nonunitary. The CCGM approximation avoids this drawback but neglects the principal part appearing in the t-matrix. Apparently, for very corrugated surfaces and at high energies, the results it yields are not of sufficient accuracy [36].

Alternative methods, closely related to this one, have been developed by Wolfe and Weare [39], and Celli, García and Hutchison [40].

IV.1.5. Rayleigh's Theory Based on a Hard Corrugated Surface Model. Application of Rayleigh's theory using a hard corrugated surface was proposed in gas-surface quantum scattering by Garibaldi *et al.* [26]. They shown rainbow and diffraction are, in a sense, one and the same phenomenon. According to their work, the rainbow pattern is the envelope of the diffraction peak intensities. Although rainbow effect was well known in optics and molecular scattering, surface rainbow, a pair of strong maxima in the scattering probability as a function of scattering angle was discovered experimentally by Smith *et al.* [41] and described theoretically by McClure [42].

Different ways of calculating diffraction intensities have been proposed. But, in any case, it seems to be widely accepted that results arising from these calculations are very useful in order to determine soft wall potentials. [13a] gives a good report of all rigorous treatments and refinemets developed with this surface model. For example, the first and most directly improvement can be obtained by allowing for the presence of an attractive well introducing bound states (SAR) into the system.

IV.1.6. The Complex Coordinate Method. This method can be viewed as an alternative for calculating complex poles of the S-matrix [43]. Resonances are treated as bound states, doing attractive this way. However, in this theory, operators can not be hermitians since we are interested in complex eigenvalues. In general, it is habitual to treat with real distance variables but, mathematically, it is possible to extend these integration variables to the complex field. Only observables have physical meaning in quantum theory (eigenvalues, etc.). Thus, if ρ is a real integration variable, we put $z = \rho\, e^{i\theta}$ corresponding to a rotation of this coordinate; here θ is the rotation angle measured from the positive real axis. Evidently, final results must be independent on θ. As a consequence, the wave vector must be a complex number ($k = \kappa\, e^{-i\beta}$ with $0 < \beta < \pi/2$ and $\kappa > 0$) if k is a pole of the S-matrix. Discrete ($k = i\kappa$) and virtual ($k = -i\kappa$) states appear also in the theory.

In atomic and molecular physics, this method has been widely used for studying resonance processes. In particular, Gerber and Moiseyev [44] applied it firstly to compute lifetimes in gas-surface scattering.

IV.1.7. Semiclassical and Classical Path Methods. Again, these methods were widely applied to gas-phase scattering problems by W.H. Miller and R.A. Marcus. Doll [45], Masel [46], and McCann and Celli [47] adapted the semiclassical theory to diffraction scattering.

Nevertheless, some computational difficulties are involved [36], due to problems in the computation of oscillatory functions via Monte Carlo methods.

A different point of view, within a semiclassical framework, were adopted by Grote and De Pristo [48] and Meyer and Toennies [49]. In this method, the vibrations of the surface are treated classically via the generalized Langevin techniques [50].

IV.1.8. Classical Formulation.   Many applications of classical mechanics to weakly bound molecular systems in order to study resonance processes can be found in the literature [51].

In gas-surface scattering, classical rainbows were found by McClure [42]. This phenomenon occurs when the Jacobian connecting scattering angles (θ, ϕ) with impact parameters cancels out. In this situation, the scattered intensity, as a function of the scattering angle, presents infinite peaks. Resonance phenomena can be also treated by solving the classical Hamilton equations [52]. If lattice vibrations are included, classical calculations become more complicated and expensive [53, 54]. A well-known handicap of these calculations is the requirement of many long-lived trajectory calculations if a trapping/desorption process occur together with the presence of some high frequency in the system (gas particle + solid surface) analyzed. In this case, the generalized Langevin equaiic n is to be prefered [50]. This method is widely applied to inelastic scattering.

Another way of tackling this problem has been the "hard cubes model" together with impulsive collision models [13b].

IV.2. TIME DEPENDENT FORMALISM

Again, the different time dependent method, which we are going to mention now, have been largely developed in gas-phase dynamics. Applications to gas-surface scattering date from 1983.

IV.2.1. Semiclassical Gaussian Wavepacket Methods.   This approach was originally due to Heller. Drolshagen and Heller [55] applied it to surface scattering. The initial state is taken as a linear combination of gaussian wavepackets and the potential $V(\underline{r})$ is expanded in powers of $(\underline{r} - \underline{r}_t)$ where $\underline{r}_t$ is the position of the center of the wavepackets at any time t. The gaussian parameters satisfy classical equations. As far as we know, this method fails for describing resonance scattering and tunneling effects.

IV.2.2. Fast Fourier Transform Method.   Firstly introduced by R. Kosloff and D. Kosloff [56] in this context, this method seeks a numerically exact solution of the time-dependent Schrödinger equation. For each time step, evaluation of the second spacial derivative of this wave function is made with the fast Fourier transform algorithm. The wave function is transformed to $\underline{k}$−space since, in this space, the action of the second spatial derivative is just the multiplication by k^2 and, after that, again this algorithm transforms back the result to $\underline{r}$−space. It is well known the numerical efficiency of this algorithm. Errors in time propagation can be overcome by expanding the evolution operator [56c]. In this method, only one column of the S-matrix, $S_{\underline{G}0}$, is calculated (where all physical interest is found). This method has been applied to resonance scattering and scattering from disorded, nonperiodic surfaces and when phonon effects are included. Propagation methods offer a complementary and valuable physical insight of the scattering process of that given by time-independent methods.

IV.2.3. Time-dependent Golden Rule. Recently, an example of slow intramolecular decay, appearing in the vibrational predissociation of vdW molecules, has been studied by using a time-dependent Golden-Rule [57]. We extended this method to the elastic scattering of ^4He atoms by the Cu(110) surface [58]. The desorption process is analyzed by means of time-dependent wavepacket techniques. SAR widths are obtained by the Fourier transform, as a function of scattering geometry parameters, of a time dependent autocorrelation function. It can be shown [57] that

$$\Gamma_{\underline{\rho}\nu}(\varepsilon) = \frac{1}{2\hbar} \int_{-\infty}^{+\infty} dt \, e^{i\,\varepsilon t/\hbar} \langle \Phi_Q(0) \mid \Phi_Q(t)\rangle \tag{IV.12}$$

with

$$\mid \Phi_Q(0)\rangle = Q \mid \Phi(0)\rangle = QV_{\underline{\rho}} \mid \underline{\rho}\nu\rangle$$

$$\mid \Phi_Q(t)\rangle = e^{-iH_o t/\hbar} \mid \Phi_Q(0)\rangle \tag{IV.13}$$

H_0 being the hamiltonian acting on the final subspace and Q the projector on its continuum part. The wavepacket at time zero was taken as the product of the quasi-bound wave function and the coupling inducing desorption. The time dependent Schrödinger equation was solved using a finite difference method proposed by Brito *et al.* [59]. This method turned out to be numerically very efficient and a good alternative to study very slow diffraction decays.

Up to now, nothing has been said about the temperature effect of the crystal surface in these two formalisms. Most of theoretical treatments of gas-surface scattering in both formalisms have been performed without an explicit temperature dependence in its dynamics. An alternative has been exposed, in the last Section, by using modified model potentials [11]. If a rigid and perfectly periodic surface model is considered, the scattering is purely elastic. For a crystal at a finite temperature, this model fails because the surface is not perfectly periodic due to the thermal motions and the probability of inelastic scattering increases. The DW theory was initially applied to explain attenuation of diffracted intensities of X-rays by a surface. This theory was extended to gas-surface scattering with a generally accepted success. As was already mentioned, it is the DW factor which possesses all information about the solid surface. This theory allow us to calculate only diffracted intensities for a perfect rigid lattice surface and the corresponding ones, at a given temperature, are obtained through the DW factor. DW and inelastic effects in atom-surface scattering have been discussed by Levi and Suhl [60] and Schinke and Gerber [18]. Discussion about these interesting problems are out of the scope of this paper.

Only thermal attenuation of elastic scattering will be treated here but following closely the surface model assumed by Lapujoulade in [17]. In particular, SAR widths and positions as a function of the temperature will be presented in the next Section.

The ^{4}He—Cu(11α) systems have been the subject of a great variety of authors, although Armand and Manson are the authors whose work is mainly addressed to these systems. Most of the methods and/or approximations exposed in this Section have been applied to these systems, except those referenced in IV.1.2, IV.1.7, IV.1.8 and IV.2.1.

V. Application to the ^{4}He—Cu(11α) (α = 0, 3, 5, 7) Systems. The Critical Kinematic Effect

Due to the arguments explicited in the two last Sections, we can say that this example is a good candidate for accomplishing a good comparison among results yielded by CC calculations and those brought on by different methods. CC results must be known whenever be possible. In our case, and owing to the model potential assumed, calculations of this type are very easy of performing, requiring few diffraction channels.

Presentation of our results will be parallel to the procedure proposed by us [33b] in order to analyze, in a convenient way, the energetic and angular regions of this scattering. Thus, the critical kinematic effect has been found out.

V.1. KINEMATIC INFORMATION

In the preceding lines, it has been stood out the peculiar characteristics of this scattering. The relative disposition of diffraction channels changes drastically according to geometric parameters chosen and, as a consequence, the manifestation of SAR. For this reason, a previous step to any dynamical calculation is to plot the diffraction channel kinetic energies versus E, θ and ϕ. These plots resume all kinematic information of the system and are, therefore, also of interest for experimentalists.

In Figs. 1a and 1b, drawings of diffraction channels and their corresponding kinetic energies as a function of θ, at 21 meV (Fig. 1a), and as a function of E, at $\theta = 60.38°$ (Fig. 1b), for the ^{4}He—Cu(113) are shown. On the left axis, specification of $\underline{G}$ vectors is made. As, in this case, the $\underline{G}$ vectors are monodimensional, $(2\pi/a)\,(m, 0)$, the pairs $(|m|, -|m|)$ represent two vectors $(2\pi/a)(|m|, 0)$ and $(2\pi/a)(-|m|, 0)$. The upper curve corresponds to $-|m|$ and the lower curve to $|m|$. The dashed lines fix the $\nu = 0$-state of $V_0(z)$ and solid lines, the thresholds. The regularity observed in Fig. 1a has been explained elsewhere [31]. Degeneration of diffraction channels (at the crossing points) in CC calculations is not a typical situation in atomic and molecular collisions. Near these points and if the order of the coupling or diffraction order, whose magnitude is given by the corresponding Fourier coefficients, is important, mixing resonant scattering [30] will predominate. Out of these crossing regions, non-resonant scattering will occur except for those regions around the dashed line or, in general, some bound state of $V_0(z)$, where resonant scattering will take place. Another advantage of this plot is that diagramatically we see which are the relevant diffraction channels in the calculations depending on what angular region we are interested, for a fixed total energy. A supplementary kinematic information is taken out from Fig. 1b. For a fixed angle, the resonant channel $\left(\rho = (2\pi/a)(m, 0)\ \text{with}\ m = 1\right)$ cuts twice to the dashed line in two different energetic regions. Notice that is the same resonance. Moreover, this secant line can be transformed into a tangent one at certain angle, critical angle [33a].

All this information is very useful for performing theoretical calculations, but also for the experimental work because a previous knowledge of these plots should serve to select adequate resonance conditions. Similar plots are valid for ^{4}He—Cu(11α) with $\alpha = 0, 5, 7$.

V.2. GOLDEN RULE (GR) RESULTS

The dynamics of the resonant phenomenon can be viewed as a half-collision process. The system is initially in a metaestable level, coupled to a family of diffraction continua through

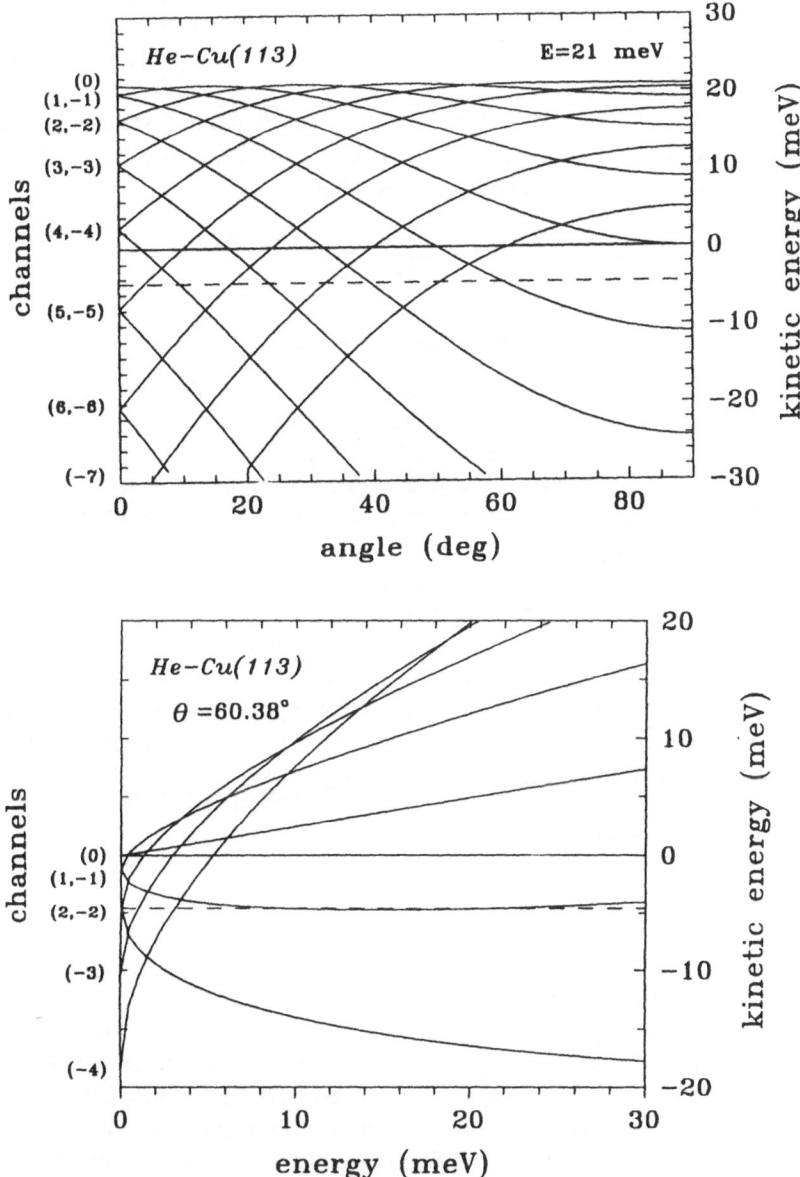

Fig. 1. Plots of diffraction channel kinetic energies against θ (Fig. 1a) and against E (Fig. 1b) for the ^{4}He—Cu(113) system. E is fixed at 21 meV in the first case and θ is fixed at 60.38° in the second one. On the left axis, specification of $\underline{G} = (2\pi/\alpha)(m,0)$-vectors is made by pairs, $(|m|, -|m|)$.

the diffracting potentials inducing the breaking up of the vdW bond (desorption process) formed between the atom and the surface.

After doing this complet kinematic study, a dynamical decoupling scheme could provide us with a global calculation of resonance features. With this, preliminary quantitative results would be known. This goal is well attained by means of the Golden Rule or related schemes. Estimates of positions and widths (or desorption rates) are very good when the corrugation of the surface is not important.

Within the simplest Golden Rule scheme, an additional benefit of using the corrugated Morse potential described in Section III, Eqs. (III.10) to (III.12), is that the widths of SAR can be expressed analytically. We have shown [58] that

$$\Gamma_{\underline{\rho}\nu}(\varepsilon) = \frac{\pi(2\gamma - 2\nu - 1)}{64\,\nu!\,\Gamma(2\gamma - \nu)} \sum_{\underline{G}} \left| \frac{V_{\underline{\rho}-\underline{G}}}{V_{\underline{0}}} \right|^2 \frac{\hbar^2\omega^2}{D} \frac{\sinh(2\pi\beta_\varepsilon)}{\cosh^2 \pi\beta_\varepsilon - \cos^2\pi(1/2 + \gamma)} \otimes$$

$$\otimes \left[(\gamma - \nu - 1/2)^2 + \beta_\varepsilon^2 + 2\gamma \right]^2 |\Gamma(1/2 + \gamma - i\beta_\varepsilon)|^2 \qquad (V.1)$$

with: $\omega = \chi(M_g/2D)^{-1/2}$, $\gamma = 2D/\hbar\omega$ and $\beta_\varepsilon = (\hbar\chi)^{-1}(2M_g\varepsilon)^{1/2}$ and $\Gamma(s)$ the gamma function of the complex number s. Here, $\underline{G}$ runs over open channels. This allow us to compare analytical and numerical results. In Table I, partial and total halfwidths issued from analytical (Eq. (V.1)) and numerical (Eq. (IV.9)) calculations, at total energy of 21 meV and for the ^{4}He—Cu(110) system, are presented. This potential sustains four levels (adsorption states or SAR) of which only the three first halfwidths have been shown in this table. The Fourier coefficients are also analytical if a cosine shape function is assumed [31]. Only contributions from the specular channel ($m = 0$) and the second open channel ($m = -1$) have been considered since the couplings for diffraction orders greater than two are negligible. The functional dependence with ε is subject to the very oscillatory behavior of the continuum wave function. Hence, a rapid decreasing of $\Gamma_{\rho\nu}$ with ε is induced. When $\pi\beta_\varepsilon \gg 1$, this dependence is practically exponential. Concerning ν levels, the halfwidth of the second one is about two times greater than those of levels $\nu = 0$ and $\nu = 2$. This fact is explained from Eq. (V.1) where the ν-dependence is explicitly known.

This analysis could be extended to different scattering geometries by calculating not only inherent widths ($\Gamma_{\rho\nu}$) but also widths associated with the experimental parameters through Eq. (IV.7). As an example, in Fig. 2, we show $\Gamma_{\rho\nu}$ (open circles) Γ_E (closed circles) and Γ_θ (crossed circles) against E (ranging from 5 to 90 meV), for the $\binom{1}{0}$ resonance of the ^{4}He—Cu(113). Each point of these curves has been obtained with the energy indicated in abscissa and the corresponding angle satisfying the kinematic condition, Eq. (IV.5). Two different scales have been used in ordinate, the left one for energetic witdhs and the right one for angular widths. The decreasing of $\Gamma_{\rho\nu}$ is governed by the so-called "energy gap law" [61]. As E increases, ε also does it, and a progressive decoupling between the two states (discret and continuum) tends to diminish the widths of the $\nu = 0$-adsorption state. Notice the two orders of magnitude of variation in all the energetic region scanned. As regards Γ_E, a maximum is observed between 10 meV and 15 meV and with a value around 4 meV. In this point of zero derivative (see Eq. (IV.7)), the stay-time should be minimum. These special resonant incidence conditions (energy and angle) have been called by us "critical conditions" [33]. This effect is the counterpart of the kinematic focussing effect [15] in inelastic scattering appearing in angular spectra. From purely kinematic arguments, this critical kinematic effect can be predicted. The critical angle is given by

$$\cos^2 \theta^* = - \left(\frac{2\pi}{a} \right)^2 \frac{\hbar^2}{2M_g\varepsilon_\nu} \qquad (V.2)$$

TABLE I

Partial and total halfwidths (in a.u.) for the $\nu = 0, 1, 2$-SAR. Analytical halfwidths are calculated from Eq. (V.1) and numerical halfwidths from Eq. (IV.9). Contributions of only two open channels are shown, $m = -1, 0$.

Resonance	Open Channels $(m, 0)$	Analytical halfwidths	Numerical halfwidths
$\binom{1}{0}0$	-1 0	.13233(-11) .47456(-7) .47457(-7)	.13233(-11) .47455(-7) .47456(-7)
$\binom{1}{1}0$	-1 0	.27945(-11) .72487(-7) .72490(-7)	.27945(-11) .72486(-7) .72489(-7)
$\binom{1}{2}0$	-1 0	.22335(-11) .48560(-7) .48563(-7)	.22335(-11) .48560(-7) .48562(-7)

The corresponding energy, E^*, is calculated from Eq. (IV.5). Another interesting aspect of this figure is that this width is always greater than $\Gamma_{\rho\nu}$. Experimentally, it is Γ_E and not $\Gamma_{\rho\nu}$ susceptible of being measured. The same is valid for Γ_θ . Again, it can be seen, at low energies, the great variation of this width with E. Contrary to the behavior of Γ_E, Γ_θ presents no maximum or minimum since, within the interval [0,90], the corresponding derivative $(d\epsilon_\rho/d\theta)_{\bar{E},\bar{\theta}}$ never goes to zero. Clearly, all this information is very useful to the experimentalists before carryng out their experiments in order to control optimum manifestations of SAR.

In dealing with more corrugated copper surfaces, it is envisaged that these widths will increase according to Eq. (V.1) where the effect of the surface roughness is ruled by the corrugation factor, $| V_G/V_0 |^2$. In fact, similar figures to Fig. 2 could be drawn for $\alpha = 5$ and 7. A comparison with the four systems ($\alpha = 0, 3, 5$ and 7) is reported in [33b].

V.3. CLOSE-COUPLING RESULTS

Within the multichannel scattering formulation, the independent variables of the S-matrix are the kinetic energies of all channels: $\{\epsilon_{\underline{G}}(E, \theta, \phi)\}_{\underline{G}}$. At the beginning of this Section, we have shown that the dynamical arrangement or configuration of all diffraction channels is quite different depending on the values taken for E, θ and ϕ ("moving threshold" multichannel scattering). In our case, E and θ are the experimental parameters to be varied,

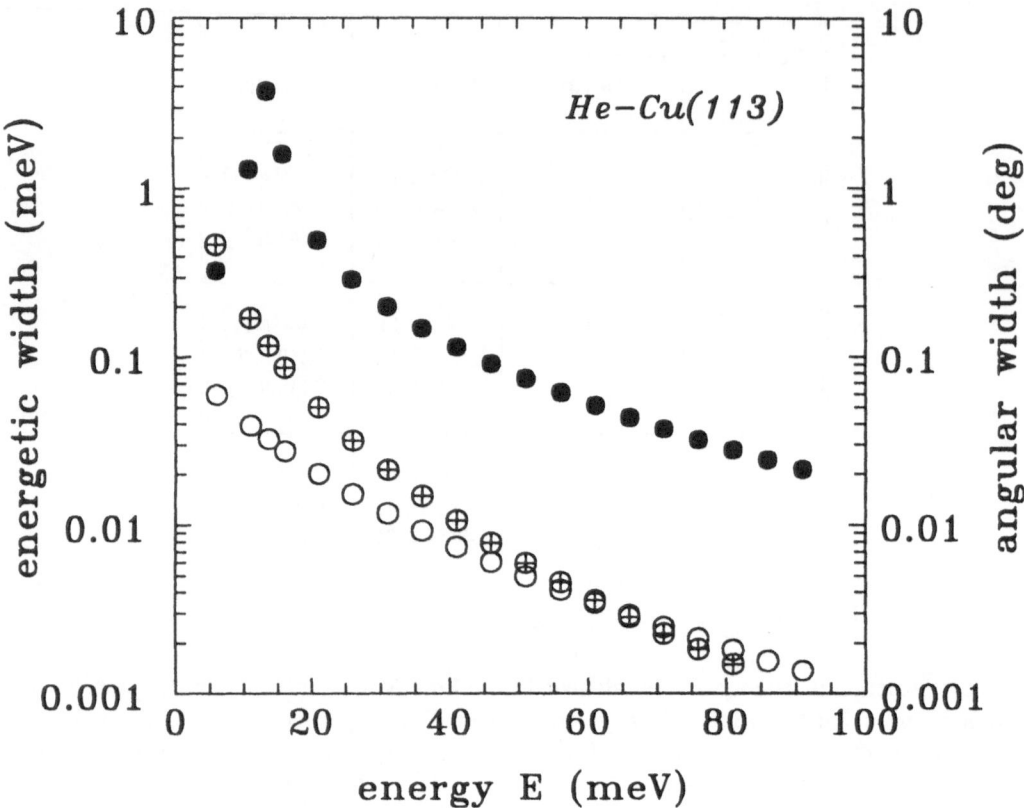

Fig. 2. $\Gamma_{\left(\begin{smallmatrix}1&0\\0&0\end{smallmatrix}\right)}$ (open circles), $\Gamma_{E\left(\begin{smallmatrix}1&0\\0&0\end{smallmatrix}\right)}$ (closed circles) and $\Gamma_{\theta\left(\begin{smallmatrix}1&0\\0&0\end{smallmatrix}\right)}$ (crossed circles) in meV and degrees, respectively, for the ^{4}He—Cu(113) system.

since the corrugation is assumed to be significant in the x-direction ($\phi = 0$). For different (E, θ)-values, extrapolation of diffracted intensities to zero temperature were made. These data were used in a fitting procedure of the potential parameters with the help of a dynamical method. The outcomes of this task, already published [17f], have served as starting point in our calculations to reproduce those diffracted intensities. At the same time, the numerical integration parameters (starting point, -5 a.u.; integration step, 0.01 a.u.; integration points, 3000) were determined for the CC algorithm in non-resonant conditions. According to Eq.

(IV.6), S and S_{bg} coincides in such conditions, due to the fact that the resonant term is practically negligible. Very often, specular and diffracted intensities are calculated in terms of their angular dependence for a total energy given. In particular, at the crossing points of Fig. 1, where certain diffraction channels are degenerated, and if the effective coupling among them is strong, the corresponding intensities can undergo important variations in magnitude.

Another source of interesting results is that supplied by resonant (critical and noncritical) scattering. We proceed now to analyze these situations together with scattering at threshold and temperature dependent resonant scattering in more detail.

V.3.1. Resonant Scattering (at non critical and critical conditions). After exploring a wide range of different resonant incidence conditions through the Golden-Rule, a local and more quantitative analysis of SAR is entailed on as a subsequent step.

Within the CC formulation, two procedures can be followed in order to calculate positions and widths of resonances. The first one is based on the study of the behavior of the specular intensity (SI), the $|S_{00}|^2$ matrix element. The second one is the so-called mixed S-matrix method [31] (MSM), requiring the evaluation of the background S-matrix, S_{bg}. Both methods are totally equivalent but for an interpretation work, the second one is more convenient. Whereas the SI method works locating an extremum value of the especular intensity (usually a maximum), the MSM method calculates particular values of two lorentzian functions depending on the position and width of the resonance studied. Thus, we can construct, from Eq. (IV.6), the two following functions

$$\mathcal{R}(\epsilon_{\underline{\rho}}) \equiv \mathcal{R}e\left\{ Tr\left[S_{bg}^+ S \right] \right\} = N - \frac{\Gamma_{\rho\nu}^2/2}{(\epsilon_{\rho} - \bar{\epsilon}_{\nu})^2 + \Gamma_{\rho\nu}^2/4}$$

$$\mathcal{L}(\epsilon_{\underline{\rho}}) \equiv \mathcal{I}m\left\{ Tr\left[S_{bg}^+ S \right] \right\} = -\frac{(\epsilon_{\rho} - \bar{\epsilon}_{\nu})\Gamma_{\rho\nu}}{(\epsilon_{\rho} - \bar{\epsilon}_{\nu})^2 + \Gamma_{\rho\nu}^2/4} \qquad (V.3)$$

provided that S and S_{bg} are unitary matrices. Here, N is the number of open channels, Tr stands for the trace and $\mathcal{R}e$ and $\mathcal{I}m$ the real and imaginary part of a complex number, respectively. With this variable, ϵ_{ρ}, the SAR is always isolated (eigenvalues of $V_0(z)$). These functions take particular values, easily deducible, for $\epsilon_{\rho} = \bar{\epsilon}_{\nu}$ and $\epsilon_{\rho} = \bar{\epsilon}_{\nu} \pm \Gamma_{\rho\nu}/2$. Deviations of the predicted behavior must be attributed to the assumption concerning S_{bg}, i.e., to be a smooth function of its argument. There is not a unique way to obtain a good S_{bg}-matrix. Among them, we can mention that proposed by Child *et al.* [62] consisting in a fitting procedure. Something similar was proposed by Schinke [25]. Two more routes can be those employed by us elsewhere [63]: to make a calculation with only open channels or with also closed channels except that sustaining the resonance. In the regime of weak coupling (weak corrugation), any of them is appropiate. Problems arise in the regime of strong coupling. A procedure working fairly well is to take into account *all* the relevant channels but keep constant the value of the kinetic energy of the resonant channel in the region out of the resonance. In doing so, all diffraction channels are allowed to move, when E or θ is varied, except the resonant channel which is fixed on an out-of-resonance position. Clearly, two requirements must be satisfied: out of the resonance region, S and S_{bg} must coincide and within this region, S_{bg} must behave smoothly. A more elaborate way could consist in making an Argand plot (imaginary part versus real part) of each S-matrix element and to fit the background contribution to [25]

$$A \exp[i(a + bq)] \tag{V.4}$$

an almost perfect circle around the origin; q being the variable of integration (E or θ in our case), and A, a and b are constants. In Figs. 3, we show Argand plots for the first element of the S (Fig. 3a) and S_{bg} (Fig. 3b) matrices between 69° and 75°. This corresponds to the $\binom{1}{0}^0$ resonance region of the ^{4}He—Cu(115) system for E = 21 meV, varying the θ angle. In Fig. 3a, the star in the third quadrant, $\theta = 72°$, is very close to the resonance position $\bar{\theta}_{CC} = 72.06°$, and with $\Gamma_{\theta}^{CC} = 1.1°$. A step size of 0.5° is used. As can be seen in Fig. 3b, the behavior of the counterpart S_{bg} element is very good when the procedure of the fixed resonant kinematic energy is chosen. At high incidence energies, this extra calculation implies more computer time since a higher number of relevant channels are involved.

If this resonance is analyzed by means of the MSM method, it can be observed the important role played by the ϵ_{ρ} variable, i.e., to be resonant variable. In fact, if two types of calculations are performed, the first one varying the θ angle at E = 21 meV (Fig. 4a), and the second one, varying E at $\theta = 72.06°$ (Fig. 4b), the same information is obtained. In both figures, the behavior predicted by Eq. (V.3) is correct in a very good approximation. The position of resonance is given by the minimum of the real part or the crossing point of the imaginary part with the zero line; and the width, by the interval between the two crossing points of the real part with the (−1)-line or the interval between the maximum and the minimum of the imaginary part. This is a very striking result because of the relevant channels involved in the calculations have relative positions radically different if E or θ is varied; in other words, the dynamical configuration changes at each value of E or θ, following the formula for ϵ_{G} (Eq. (IV.4)), moving the thresholds in a very different manner. But, in a sense, it is only the information arising from the resonant channel, ϵ_{ρ}, which gives us the main properties of the resonance. This tendency is more pronounced in the ^{4}He—Cu(110) system. Similar figures to those of Figs. 4a and 4b can be plotted. However, in Fig. 5a and 5b, we present another form of standing out the same fact. For the $\binom{1}{0}^0$ resonance, a contour map of the specular intensity (Fig. 5a) and the absolute value of the kinetic energy $\epsilon_{(1,0)}$ (Fig. 5b), as a function of E and θ, show these curves are well represented by straight lines (in the resonance region) of equal slopes. Due to the weak corrugation of this system, only the resonant and specular channels are necessary to describe properly this scattering. For other kinetic energies, these contour maps display different slopes.

Now, it would be interesting to express Eq. (IV.6) as a function of E or θ, which are the experimental parameters susceptible of being changed. The ϵ_{ρ} variable is fundamental from the theoretical point of view (internal variable), but all diffraction spectra are very often obtained with those variables. Thus, if the function ϵ_{ρ} is developed in a Taylor series around the resonance position, and retaining the two first terms, we have, after a little algebra,

$$S(q) \simeq S_{bg}(q) \left[I - \frac{i \mathcal{A}'}{(q - \bar{q}) + i \Gamma_{qv}/2} \right] \tag{V.5}$$

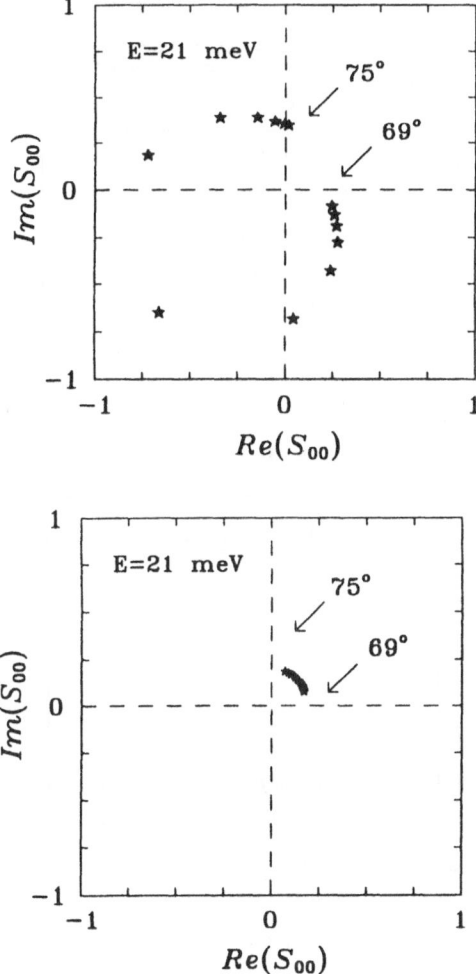

Fig. 3. For the ^{4}He—Cu(115) system, Argand plots of the first element of the S-matrix (Fig. 3a) and S_{bg}-matrix (Fig. 3b), at $E = 21$ meV, around the first resonance position, $\tilde{\theta} = 72.06°$. Step is 0.5°.

q being E or θ depending on what variable is taken in the experiment, keeping constant the other. $\mathcal{A}'$ and $\Gamma_{q\nu}$ are related to $\mathcal{A}$ and $\Gamma_{\rho\nu}$, respectively, by the Jacobian derivatives

$$\mathcal{A}' = \mathcal{A} \left(\frac{d\epsilon_\rho}{dq} \right)^{-1}_{\bar{q}}$$

$$\Gamma_{q\nu} = \Gamma_{\rho\nu} \left(\frac{d\epsilon_\rho}{dq} \right)^{-1}_{\bar{q}} \tag{V.6}$$

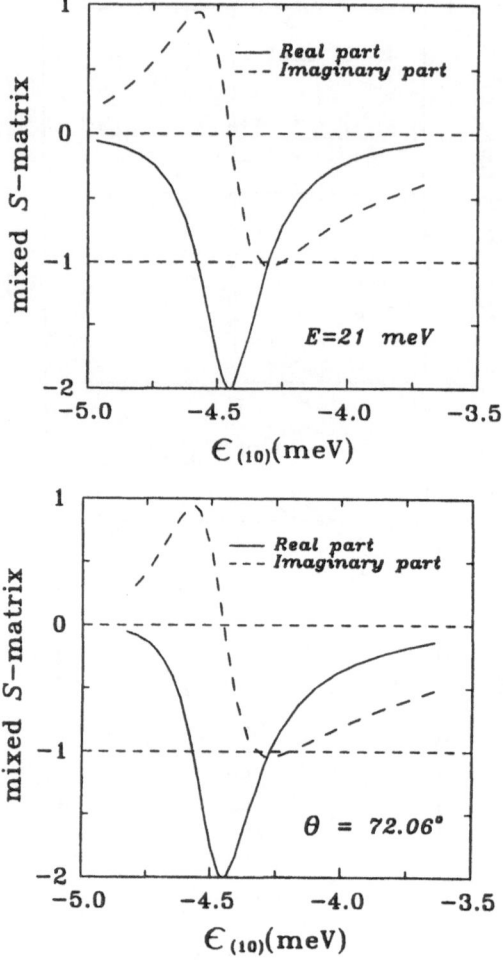

Fig. 4. θ (Fig. 4a) and E (Fig. 4b) variations in the MSM method for the first resonance of the ^{4}He—Cu(110) system.

From Eq. (V.5), it is easily deduced expressions equivalent to Eqs. (V.3),

$$\mathcal{R}(q) = N - \frac{\Gamma_{q\nu}^2/2}{(q-\bar{q})^2+\Gamma_{q\nu}^2/4}$$

$$\mathcal{L}(q) = -\frac{(q-\bar{q})\,\Gamma_{q\nu}}{(q-\bar{q})^2+\Gamma_{q\nu}^2/4}$$

(V.7)

This tranformation preserves the lorentzian forms of the $\mathcal{R}$ and $\mathcal{L}$ functions. The widths calculated by means of Eqs. (V.7) are equal to those issued from Eq. (IV.7). These widths are asymmetric and quantitatively important, specially for the $Cu(115)$ and $Cu(117)$ surfaces (several degrees or meVs). This is easily understood because the dynamical sourronding at each side of resonance position changes in a very different manner. Moreover, this situation

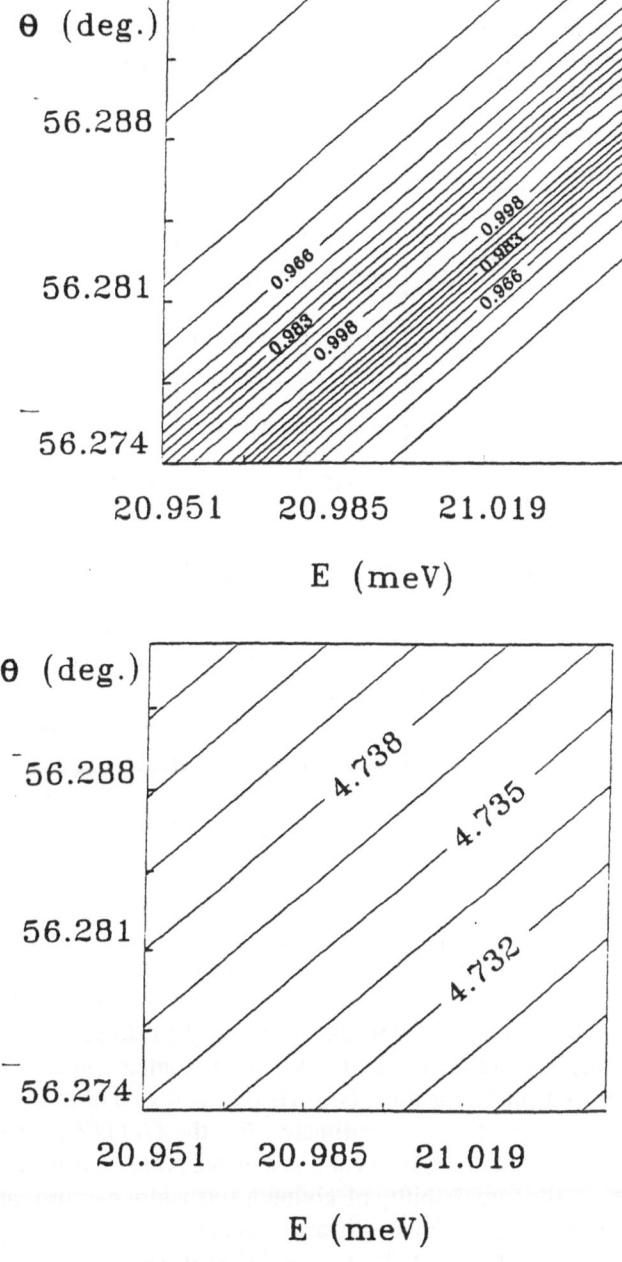

Fig. 5. Contour maps of the specular intensity (Fig. 5a) and the absolute value of the kinetic energy, $\epsilon_{(1,0)}$ (Fig. 5b), as a function of E and θ for the $\left(\begin{smallmatrix}1&0\\0&\end{smallmatrix}\right)$ resonance of the ^{4}He—Cu(110) system.

TABLE II

Positions and widths for the $\left(^{1\ 0}_{\ \ 0}\right)$ resonance at 21 meV. Golden-Rule (GR) and Close-Coupling (CC) results for the ^{4}He—Cu(11α) $(\alpha = 0, 3, 5, 7)$ systems. Energetic positions and widths in meVs, angular positions and widths in degrees.

Face	Method	$\bar{\varepsilon}$	$\bar{E}$	$\bar{\theta}$	$\Gamma_{\left(^{1\ 0}_{\ \ 0}\right)}$	Γ_E	Γ_θ
$\alpha = 0$	GR	-4.737	21.00	56.287	1.291(-3)	0.0163	5.734(-3)
	CC	-4.734	21.00	56.281	1.288(-3)	0.0162	5.719(-3)
$\alpha = 3$	GR	-4.582	21.00	60.379	0.020	0.497	0.0503
	CC	-4.567	21.00	60.342	0.01995	0.484	0.0499
$\alpha = 5$	GR	-4.582	21.00	72.585	0.315	5.993	1.301
	CC	-4.46	21.00	72.059	0.28	6.05	1.102
$\alpha = 7$	GR	-4.582	21.00	84.311	0.676	6.885	$(\Gamma_\theta/2)^- = 3.287$
	CC	-4.444	21.00	82.787	0.589	6.35	$(\Gamma_\theta/2)^- = 2.53$

induces maxima or minima in diffraction spectra. In fact, some rules have been given in order to predict these maxima or minima [39, 40]. With analogous diagrams to those of Figs. 5a and 5b but E and θ taken as independent variables, positions and widths of SAR are calculated. In Table II, GR and CC results for the $\left(^{1\ 0}_{\ \ 0}\right)$ resonance at 21 meV (one of the experimental energies) and for the four faces are presented. Only θ has been varied. It can be seen the agreement between the two sets (GR and CC) of results is really good. GR results must be considered as good estimates. For the $Cu(117)$, where four Fourier coefficients are non-negligible, the agreement is somewhat poor. Also, the minus sign found in Γ_θ is a reminder of the impossibility of giving a full width because, at $\theta = \pi/2$, the specular channel is closed. Up to eleven diffraction channels have been employed in these calculations, particulary for the $Cu(115)$ and $Cu(117)$ surfaces.

Regarding comparison with the experimental results, usually these results are reported as diffraction spectra where θ is taken as independent variable. From these plots we can extract positions and widths. In general, we can say that the agreement is fairly good. Deviations must be attributed to the model potential used. For example, for $\alpha = 5$, the experimental values are: $\bar{\varepsilon} = (4.45 \pm 0.15)$meV, $\Gamma_{\left(^{1\ 0}_{\ \ 0}\right)} = 0.35$ meV, $\bar{\theta} = 71.3°$ and $\Gamma_\theta = $

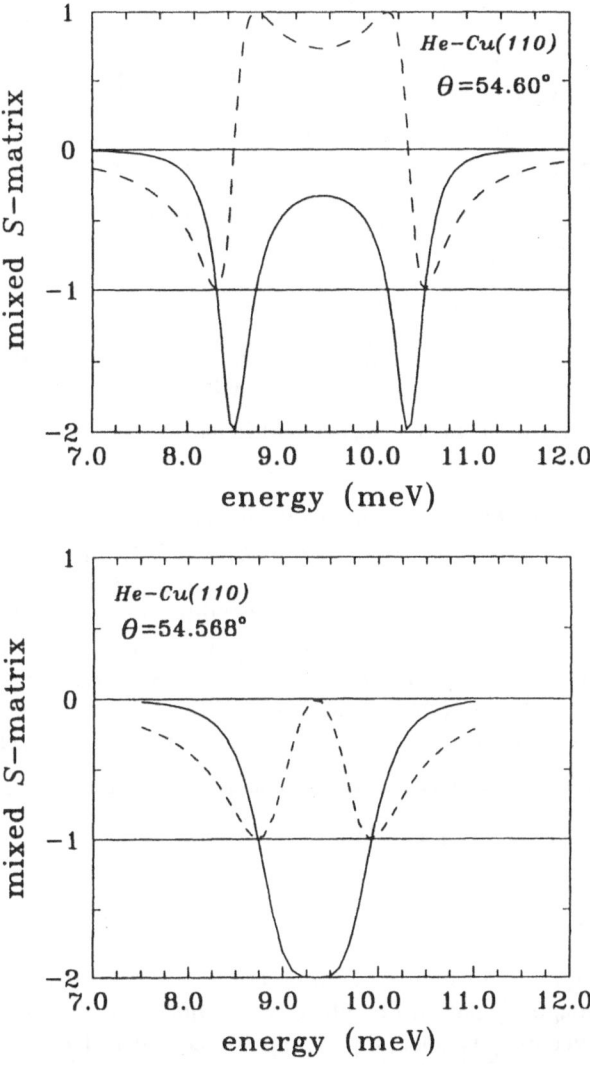

Fig. 6. $\mathcal{R}$ and $\mathcal{L}$ functions for the first resonance of the ^{4}He—Cu(110) system very near to the critical point (Fig. 6a) and at the critical point (Fig. 6b).

1.5°. Particulary, at 21meV, resonances for the Cu(110) surface have not been observed. We proposed [14] to measure angular and energetic widths at low incident energies ($< 21meV$) where the corresponding widths could be detected. The calculated energy widths have a predictive value since, for these systems, experiments varying E have not been yet accomplished.

Now, we focus attention on critical scattering. The analysis of this scattering was firstly proposed by us [14]. In these conditions, a very important enhancement of resonance

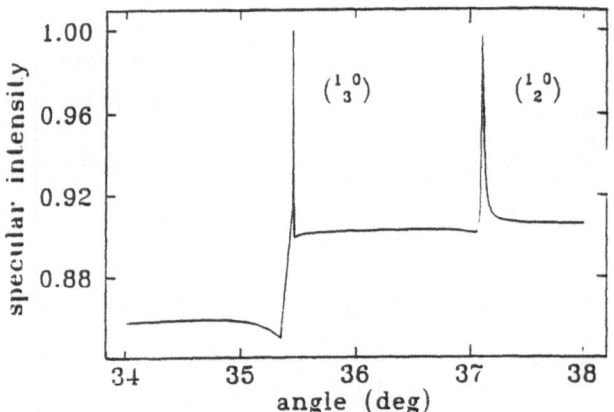

Fig. 7. For the ^{4}He—Cu(110) system, the specular intensity versus θ at 9 meV is displayed.

features occurs. In Fig. 6a and 6b, we display the behavior of the $\mathcal{R}$ and $\mathcal{L}$ funtions very near of the critical point, $\theta = 54.60°$ (Fig. 6a), and at the critical value, $\theta = 54.568°$ (Fig. 6b), for the $\binom{1\ 0}{0}$ resonance of the ^{4}He—Cu(110) system. It can be noticed how the same resonance is manifested at two different conditions of E and θ and, after very little variations of E and θ, the two peaks come together to produce a unique peak. A remarkable loss of the lorentzian behavior is found. Eq. (V.7) is no longer adequate. The energetic derivative is zero and new terms of the Taylor series must be taken into acount. If the second derivative is retained, Eq. (V.7) is transformed to (energetic version)

$$\mathcal{R}(E) = N - \frac{\Gamma_{E\nu}^4/8}{(E - \bar{E})^4 + \Gamma_{E\nu}^4/16}$$

$$\mathcal{L}(E) = -\frac{(E - \bar{E})^2\,\Gamma_{E\nu}^2/2}{(E - \bar{E})^4 + \Gamma_{E\nu}^4/16} \tag{V.8}$$

The angular version is not valid here because of the θ-derivative never is zero. The shapes of the real and imaginary part of Fig. 6b are well adjusted by Eq. (V.8). Nevertheless, if a plot of $\mathcal{R}$ and $\mathcal{L}$ versus ϵ_ρ is made, Eq. (V.3) is still adequate. The $\Gamma_{\rho\nu}$ and $\Gamma_{E\nu}$ calculated for this resonance, at these critical conditions, are practically equal to those estimated by the Golden Rule. We expect that this scattering will be analyzed experimentally in the very near future.

V.3.2. Scattering at Threshold. Threshold resonances are a characteristic feature of diffractive systems being, specially, sensitive to the details of the interaction potential. Although they are not true resonances, their manifestations are very similar to those corresponding to SAR. Following to Armand and Mason [64], when the kinetic energy of some open diffraction channel tends to zero, its Jacobian angular derivative diverges and, consequently, the variation of its associated diffraction intensity with θ also does it, unless this intensity goes to zero at least as fast as the kinetic energy. Experimental detection of these resonances is not easy within the constraints of current experiments.

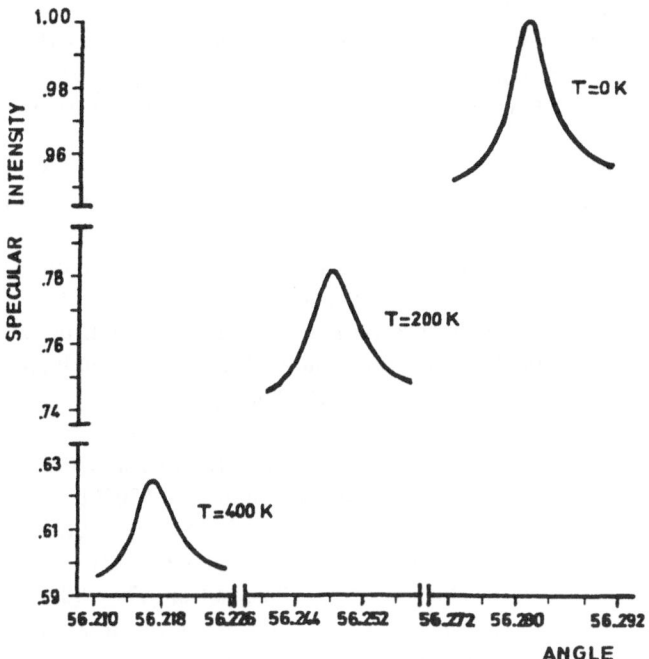

Fig. 8. For the ^{4}He—Cu(110) system, its first resonance is plotted for three different temperatures: 0, 200 and 400 K.

From the theoretical point of view, calculations performed at threshold are not difficult. In Fig. 7, and for the system ^{4}He/Cu(110), we present the specular intensity versus the incidence angle, at 9 meV. This angular region corresponds to the closure of the $\rho = (1,0)$-channel. The shape of this curve is easily understood: we observe a decreasing of the specular intensity just before the closure of the $(1,0)$-channel (threshold region) at a threshold angle but, accidentally, to the same angle practically the $\left(^1_3{}^0\right)$-resonance takes place. This effect (sudden variation) is enhanced at low incident energies where all the relevant channels are much closer than for 21 meV and, therefore, the effective couplings among them are stronger. Hence, for more corrugated surfaces, this effect could be observed.

V.3.3. Temperature Dependent Resonant Scattering. The simplest way (and more usual) for describing the thermal attenuation of diffraction intensities is by means of the well-known DW factor. The attenuated intensities are expressed by the following formula

$$I_{\underline{G}}(T) = I_{\underline{G}}(0) \exp[-2W_{\underline{G}}(T)] \qquad (V.9)$$

where $I_{\underline{G}}(0)$ is the $\underline{G}$-diffracted intensity at 0 K and the argument of the DW factor can be written as

$$2W_{\underline{G}}(T) = | \Delta\underline{K}_{\underline{G}})|^2 < u_z^2 >_T \qquad (V.10)$$

The first factor of this product indicates the normal momentum transfer during the collision. For $< u_z^2 >_T$, we have taken the values given by Lapujoulade *et al.* [17d] and issued from a fitting with the experimental data.

For the ^{4}He/Cu(110) system and according to Eqs. (III.13) and (III.14) (the model potential modified by the temperature effect), the potential parameters do not undergo important changes when temperature increases from 0 to 770 K (experimental range of variation). The same tendency is observed for the effective couplings. Accordingly, the dynamical results obtained with this modified potential are very similar to those calculated for $T = 0$ K. Even more, at $T = 0$ K, the specular intensity amounts to 0.7694 (with $\theta = 20°$ and $E = 21$ meV) and at $T = 773$ K, we have 0.7753. Thus, the process becomes more specular, decreasing the other diffraction intensities in order to conserve the unitarity of the S-matrix. Therefore, we must conclude attenuation of diffracted intensities does not stem from the temperature dependence on potential but only from the DW factor. New developments of attenuation theory are being studied. See, for example, [65].

Analogously, the same is found at resonant conditions. Changes in positions of resonances are rather appreciable theoretically. Variations of $0.03°$ each $200K$ in angular positions have been calculated by us. Notice that this variation is five times greater than the angular width ($\sim 0.0057°$) of the $\left(^1_0{}^0\right)$ resonance of the ^{4}He/Cu(110) system, at 21 meV and $T = 0K$. By using the SI method and Eq. (V.9), this SAR is plotted in Fig. 8 at three different temperatures: $0, 200$ and $400K$. Three scales are necessary due to the attenuation of the specular intensity and the shift of the angular position. The widths are very similar. This result is well known experimentally [17a]. In fact, these small changes in Γ_θ and Γ_E are negligible within the experimental resolution limits. Maybe, at critical conditions, resonance features are also enhanced.

References

[1] F.O. Goodman and H.Y. Wachman: in *Dynamics of Gas-Surface Scattring*, Academic Press, Inc., N.Y., 1976. In Chapter 1, it is presented a very interesting historical development of experiment and theory in this field.

[2] O. Stern and his group: *Zeits. f. Physik* **53** (1929) 766; ibid., **61** (1930) 95; ibid., **73** (1931) 348.

[3] J.E. Lennard-Jones and F.A. Devonshire: *Nature* **137** (1936) 1069; also a series of eight papers in *Proc. Roy. Soc.* **A150** (1935) 442 and 456; ibid., **A156** (1936) 6, 29 and 37; ibid., **A158** (1937) 242, 253 and 269. Some of them are signed by C. Strachan.

[4] For a review, J.P. Estrup, and E.G. McRae: *Surf. Sci.* **25** (1971) 1.

[5] a) A. Pentero: *Catal. Rev.* **5** (1971) 199;
b) C.R. Brundle: *J. Vac. Sci. Technol.* **11** (1974) 212.
For other surface spectroscopies see, for example, the following books:
c) E. Drunglis, R.D. Gretz, and R.I. Jaffee: *Molecular Processes on Solid Surfaces*, McGraw-Hill Book Company, N.Y., (1968);
d) E. Drauglis and R.I. Jaffee: *The Physical Basis for Heterogeneous Catalysis*, Plenum Press, N.Y., (1976).

[6] a) J.N. Smith, Jr.: *Surf. Sci.* **34** (1973) 613;
b) J.P. Toennies: *Appl. Phys.* **3** (1974) 91.

[7] a) N. Cabrera: **Disc. Faraday Soc.** **28** (1959) 16;
b) R.W. Zwanzig: *J. Chem. Phys.* **32** (1960) 1173;
c) F.O. Goodman: *J. Phys. Chem. Solids* **23** (1962) 1269;
d) C.M. Chambers and E.T. Kinzer: *Surf. Sci.* **4** (1966) 33.

[8] a) R.A. Oman: *J. Chem. Phys.* **48** (1968) 3919;
b) J.D. McClure: *J. Chem. Phys.* **52** (1970) 2712.

[9] a) R.M. Logan and R.E. Stickney: *J. Chem. Phys.* **44** (1966) 195;
b) R.M. Logan, and J.C. Keck: *J. Chem. Phys.* **49** (1968) 860.

[10] A. Tsuchida: *Surf. Sci.* **14** (1969) 375.

[11] N. Cabrera, V. Celli, F.O. Goodman, and J.R. Manson: *Surf. Sci.* **19** (1970) 67.

[12] a) G. Wolken, Jr.: *J. Chem. Phys.* **58** (1973) 3047;
     b) for a review, G. Wolken, Jr.: in *Dynamics of Molecular Collisions*, Part A, edited by W.H. Miller, Plenum Press, N.Y., (1976), p. 211.
[13] a) H. Hoinkes: *Rev. Mod. Phys.* **52** (1980) 933;
     b) J.A. Barker and D.J. Auerbach: *Surf. Sci. Reports* **4** (1985) 1.
[14] M. Hernández, S. Miret-Artés, P. Villarreal, and G. Delgado-Barrio: *Surf. Sci.* (in press).
[15] G. Benedek: *Phys. Rev. Lett.* **35** (1975) 234.
[16] a) G. Benedek and U. Valbusa, (Eds.): *Dynamics of Gas-Surface Interaction*, Springer Series in Chemical Physics 21 (Springer, Berlin, 1982);
     b) A.E. De Pristo and A.Kara: *Adv. Chem. Phys.* **77** (1989) 163.
[17] a) J. Lapujoulade and Y. Lejay: *J. Chem. Phys.* **63** (1975) 1389;
     b) J. Perrau and J. Lapujoulade: *Surf. Sci.* **119** (1982) L292; ibid., **122** (1982) 341;
     c) B. Salanon, G. Armand, J. Perreau, and J. Lapujoulade: *Surf. Sci.* **127** (1983) 135;
     d) J. Lapujoulade, J. Perreau, and A. Kara: **Surf. Sci. 129** (1983) 59;
     e) D. Gorse and J. Lapujoulade: *Surf. Sci.* **162** (1985) 847);
     f) D. Gorse, B. Salanon, F. Fabre, A. Kara, J. Perreau, G. Armand, and J. Lapujoulade: *Surf. Sci.* **147** (1984) 611.
[18] R. Schinke and R.B. Gerber: *J. Chem. Phys.* **82** (1985) 1567.
[19] a) D. Eichenauer and J.P. Toennies: *J. Chem. Phys.* **85** (1986) 532;
     b) D. Eichenauer, U. Harten, J.P. Toennies, and V. Celli: *J. Chem. Phys.* **86** (1987) 3693.
[20] W.A. Steele: *The Interaction of Gases with Solid Surface*, Pergamon, Oxford, 1974.
[21] J.D. Hirschfelder, C.F. Curtiss, and R.B. Bird: *Molecular Theory of Gases and Liquids*, Wiley, N.Y., 1954.
[22] I.P. Batra: in *Topics in Current Physics* **29** (1982).
[23] N. Esbjerg and J. Nørskov: *Phys. Rev. Lett.* **45** (1980) 807.
[24] D.R. Hamann: *Phys. Rev. Lett.* **46** (1981) 1227.
[25] R. Schinke: *Surf. Sci.* **127** (1983) 283.
[26] U. Garibaldi, A.C. Levi, R. Spadacini, and G.E. Tommei: *Surf. Sci.* **48** (1975) 649.
[27] J.S. Hutchison: *Phys. Rev.* **B22** (1980) 5671.
[28] G. Armand and J.R. Manson: *Phys. Rev. Lett.* **43** (1979) 1839.
[29] G. Armand and J.R. Manson: *Surf. Sci.* **119** (1982) L299.
[30] a) H. Chow and E.D. Thompson: *Surf. Sci.* **82** (1979) 1;
     b) H. Chow: *Surf. Sci.* **62** (1977) 487.
[31] M. Hernández, O. Roncero, S. Miret-Artés, P. Villarreal, and G. Delgado-Barrio: *J. Chem. Phys.* **90** (1989) 3823.
[32] J.R. Taylor: *Scattering Theory*, Wiley, N.Y., 1972, p. 411.
[33] M. Hernández, S. Miret-Artés, P. Villarreal, and G. Delgado-Barrio: *Surf. Sci.* (in press);
     b) ibid., *Surf. Sci.* (submitted).
[34] G. Benedek: *Phys. Rev. Lett.* **35** (1975) 234.
[35] R.B. Gerber, A.T. Yinnon, and J.N. Murrel: *Chem. Phys.* **31** (1978) 1; ibid., **33** (1978) 131.
[36] R.B. Gerber: *Chem. Rev.* **87** (1987) 29.
[37] S. Miret-Artés, O. Roncero, P. Villarreal,and G. Delgado-Barrio,: *J. Phys. Chem.* **91** (1987) 5623.
[38] M. Gell-Mann and M.L. Goldberger: *Phys. Rev.* **91** (1953) 398.
[39] K. Wolfe and J. Weare: *Phys. Rev. Lett.* **41** (1978) 1663.
[40] V. Celli, N. García, and J. Hutchison: *Surf. Sci.* **87** (1979) 112.
[41] J.N. Smith, Jr., D.R. O'Keefe, H. Saltsburg, and R.L. Palmer: *J. Chem. Phys.* **50** (1969) 4667.
[42] J.D. McClure: *J.Chem. Phys.* **52** (1970) 2712; **57** (1972) 2810, 2823.
[43] B. Simon: *Ann. Math.* **97** (1973) 247.
[44] a) N. Moiseyev, T. Maniv, R. Elber, and R.B. Gerber: *Mol. Phys.* **55** (1985) 1369;
     b) T. Maniv, E. Engdhal, and N. Moiseyev: *J. Chem. Phys.* **86** (1987) 1048; **88** (1988) 5864;
     c) E. Engdahl, T. Maniv, and N. Moiseyev: submitted to *Phys. Rev. B.*
[45] J.D. Doll: *Chem. Phys.* **3** (1974) 257; *J. Chem. Phys.* **61** (1974) 954.
[46] R.I. Masel, R.P. Merrill, and W.H. Miller: *J. Chem. Phys.* **64** (1976) 45.
[47] K.J. McCann and V. Celli: *Surf. Sci.* **61** (1976) 10.
[48] R.F. Grote and A.E. De Pristo: *Surf. Sci.* **131** (1983) 491.
[49] H.D. Meyer and J.P. Toennies: *Surf. Sci.* **148** (1984) 58.
[50] a) S.A. Adellman and J.D. Doll: *J. Chem. Phys.* **64** (1976) 2375;
     b) J.C. Tully: *J. Chem. Phys.* **73** (1980) 1975; *Surf. Sci.* **111** (1981) 461.
[51] For example, see G. Delgado-Barrio,P. Villarreal,P. Mareca, and G. Albelda: *J. Chem. Phys.* **78** (1983) 280.

[52] N. Moiseyev, T. Maniv, R. Elber, and R.B. Gerber: *Mol. Phys.* **55** (1985) 1369.

[53] R.W. Hockney and J.W. Eastwood: *Computer Simulation using Particles*, McGraw-Hill, N.Y., 1981.

[54] J.A. Barker, D.R. Dion, and R.P. Merrill: *Surf. Sci.* **95** (1980) 15.

[55] G. Drolshagen and E.J. Heller: *J. Chem. Phys.* **79** (1983) 2072.

[56] **a)** R. Kosloff and D. Kosloff: *J. Chem. Phys.* **79** (1983) 1823;
      b) A.T. Yinnon and R. Kosloff: *Chem. Phys. Lett.* **102** (1983) 216;
      c) H. Tal-Ezer and R. Kosloff: *J. Chem. Phys.* **81** (1984) 367.

[57] P. Villarreal,S. Miret-Artés,O. Roncero,G. Delgado-Barrio,J.A. Beswick, N. Halberstadt, and R.D. Coalson: *J. Chem. Phys.* (in press).

[58] M. Hernández,S. Serna O. Roncero,S. Miret-Artés,, P. Villarreal and G. Delgado-Barrio,: *Surf. Sci.* (in press).

[59] R. Brito, J.A. Cuesta, and A.F. Rañada: *Phys. Lett.* **A128** (1988) 360.

[60] **a)** A.C. Levi and H. Suhl: *Surf. Sci.* **88** (1979) 221;
      b) A.C. Levi: *Nuovo Cimento* **54B** (1979) 357.

[61] J.A. Beswick and J. Jortner: *Adv. Chem. Phys.* **47** (1981) 363.

[62] C.J. Ashton, M.S. Child, and J.M. Hutson: *J. Chem. Phys.* **78** (1983) 4025.

[63] G. Delgado-Barrio,P. Villarreal,P. Mareca, and J.A. Beswick: *Int. J. Quantum Chem.* **27** (1985) 173.

[64] G. Armand and J.R. Manson: *Surf. Sci.* **169** (1986) 216.

[65] J.R. Manson and G. Armand: *Surf. Sci.* **184** (1987) 511.

LARGE PREDICTION OF RATE CONSTANTS OF POLYATOMIC REACTIONS: PERMANENT CHALLENGE TO THEORETICAL CHEMISTS

L. ZÜLICKE
Zentralinstitut für physikalische Chemie
Rudower Chaussee 5
Berlin-Adlershof
D-O-1199
Germany

ABSTRACT. A consequent theoretical description of the dynamics of elementary chemical processes is practicable, using the available standard methodology, only for the most simple systems (e.g. exchange reactions of an atom with a diatomic molecule). This paper gives, after posing the problem, an outline of a general theoretical scheme for treating reactions in polyatomic systems based on the reaction path concept and on a dimensional reduction of the nulear dynamical problem, thus indicating possible extensions of the range of applicability of the theory. Specific aspects of the procedure are illustrated by examples.

1. Introduction

The subject of the present study is a (macroscopic) elementary chemical reaction, for example a bimolecular one:

$$X_1 + X_2 \rightarrow Y_1 + Y_2 + \dots \tag{1.1}$$

where $X_1, X_2, Y_1, Y_2, \dots$ are chemical substances (the dots include the case of more than two product species). The reaction rate equation for thermal equilibrium is given by:

$$d[Y_1]/dt = k(T)[X_1][X_2] \tag{1.2}$$

(square brackets denote here concentrations), and the rate coefficient (T) obeys frequently the Arrhenius law:

$$k(T) = A \exp(-E_a/RT) \tag{1.3}$$

(Arrhenius, 1889) where R denotes the gas constant and T the temperature. The parameters E_a and A, the activation energy and the pre-exponential factor, respectively, are supposed to be independent of temperature.

In the more than one hundred years since the formulation of the Arrhenius law, much trouble has been taken to interpret its physical content, to analyse it on the molecular level and to make predictions on this basis. A very important benchmark of this development was

L. A. Montero and Y. G. Smeyers (eds.), Trends in Applied Theoretical Chemistry, 51–69, 1992.
© 1992 *Kluwer Academic Publishers.*

the Transition State Theory as developed by Eyring, Wigner and others [1, 2]. It remained for several decades the main instrument of theoretical reaction kinetics in spite of its severe restrictions like the validity for relatively small temperature ranges only and the premise of thermal equilibrium.

A new approach to the theoretical fundament of reaction kinetics came up in the sixtieth when the development of molecular beam techniques and laser techniques made it possible to study atomic and molecular interactions under single-collision conditions (for references see, e.g., [3]) thus enabling to separate the dynamical from the statistical problem. This new field, now frequently called 'reaction dynamics', opened a very fruitful era of research, both experimental and theoretical. Nevertheless, particularly the theoretical treatment is still in its starting period because of the tremendous difficulties which arise in the investigation of real systems, except only for the most simple cases. This is the problem to be discussed in the present paper.

2. Formulation of the Problem

2.1. BASIC CONCEPTS

Macroscopic elementary reactions result from a complicated interplay of various types of microscopic (molecular) interaction processes. In the case of the bimolecular reaction (1.1) they can be formulated as

$$X_1(i) + X_2(j) \rightarrow Y_1(l) + Y_2(m) + \ldots \tag{2.1}$$

where $X_1, X_2, Y_1, Y_2, \ldots$ now denote atomic or molecular species, the letters i, j, l, m symbolize 'collective quantum numbers' which characterize the electronic, vibrational and rotational states of the respective species.

An elementary process like (2.1) can be experimentally realized in a crossed molecular beam apparatus. If $\mathcal{N}$ is the number of product particles Y_1(in state l), the fraction of it registered per second in the detector with aperture dF placed at distance R and angular position ϑ(polar angle), φ(azimuth) with respect to the interaction volume $\mathcal{V}$ where the two beams of reactant particles $X_1(i)$ and $X_2(j)$ cross with relative velocity u, gives the rate dr of process (2.1):

$$dr \equiv d^2\mathcal{N}\{Y_1(l)\}/dt = g(ij \mid u \mid lm \parallel \Omega)u\mathcal{V}d\Omega[X_1(i)][X_2(j)] \tag{2.2}$$

where $\Omega \equiv (\vartheta, \varphi)$ and $d\Omega = dF/R^2$, the solid angle element corresponding to dF. The proportionality factor $g(ij \mid u \mid lm \parallel \Omega)$ is called the *differential cross-section* of the process under consideration. Integration over all directions ϑ, φ gives the *integral cross-section*

$$\sigma(ij \mid u \mid lm) \equiv \int d\Omega g(ij \mid u \mid lm \parallel \Omega), \tag{2.3}$$

and partial or complete summation and statistical averaging of σ leads to the macroscopic *rate coefficient* $k(T)$ which depends, in the case of complete thermal equilibrium, only on temperature T.

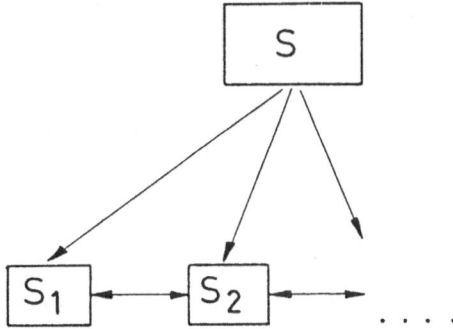

Fig. 1. Partitioning of a system S into subsystems $S_1, S_2, \ldots$.

2.2. THEORETICAL DESCRIPTION

The theoretical treatment of an elementary molecular interaction process like (2.1) involves the solution of the dynamical problem of N^e electrons and N^n nuclei constituting the system $\{X_1, X_2\}$ in which the process takes place. Since a dynamical problem becomes very complicated as soon as the number of variables in the equations of motion (i.e. the number of degrees of freedom of the system) exceeds very small values, already the simplest real systems require simplifications and approximations, and for somewhat more complex, chemically interesting systems the theoretical treatment seems almost hopeless.

The general strategy of the theory is to build up a hierarchy of 'dimensional reductions', namely to partition the systems S into two (ore more) subsystems $S_1, S_2, \ldots$ which are dynamically simpler (see Figure 1), to treat their dynamics separately and to take the coupling between the subsystems into account in an approximate way. Some stages of this hierarchy are well known and more or less standard: the (exact) separation of the overall translation (and rotation) of a molecular system, the Born-Oppenheimer separation of electronic and nuclear motions, and the independent-electron (Hartree-Fock) model. Likewise such a separation procedure should frequently be possible for the nuclear motions, not only in the case of vibrational motions of a stable molecule (normal vibrations) but also for the dynamics of a molecular rearrangement process.

Let us briefly recapitulate the usual three-step scheme [4] based on the Born-Oppenheimer separation, the starting point being the nonrelativistic approximation. The N^e electrons are described by position vectors $\xi_1, \xi_2, \ldots, \xi_{N^e}$ which we collect in a vector $\xi \equiv \{\xi_1, \ldots\}$, analogously for the N^n nuclei, $\mathbf{x} \equiv \{\mathbf{x}_1, \ldots\}$. The Hamiltonian for the whole system of N^e electrons and N^n nuclei can be written as

$$\hat{H} = \hat{T}^n + \hat{T}^e + V^{ne}(\mathbf{x}, \xi) + V^{ee}(\xi) + V^{nn}(\mathbf{x}), \tag{2.4}$$

the part

$$\hat{H}^e = \hat{T}^e + V^{ne}(\mathbf{x}, \xi) + V^{ee}(\xi), \tag{2.5}$$

the so-called electronic Hamiltonian, governs the motion of the electrons. Now we can formulate the three-step scheme mentioned above:

(I) Dynamics of the electronic motion, taking the nuclear coordinates $\mathbf{x}$ fixed:

$$\hat{H}^e \phi_k(\xi \mid \mathbf{x}) = E_k^e(\mathbf{x})\phi_k(\xi \mid \mathbf{x}). \tag{2.6}$$

(II) Dynamics of the nuclear motion, after expanding the total electronic-nuclear wavefunction $\Xi(\mathbf{x}, \xi)$ in the basis of the electronic wavefunctions ϕ_k,

$$\Xi(\mathbf{x}, \xi) = \sum_{k'} \phi_{k'}(\xi \mid \mathbf{x})\Psi_{k'}(\mathbf{x}), \tag{2.7}$$

and inserting into the Schrödinger equation with the total Hamiltonian $\hat{H}$, Equation (2.4):

$$\{\hat{T}^n + U_k(\mathbf{x})\}\Psi_k(\mathbf{x}) + \sum_{k'} \hat{C}_{kk'}\Psi_{k'} = 0, \quad (k = 0, 1, \ldots) \tag{2.8}$$

where the potential energy function is approximated by

$$U_k(\mathbf{x}) = E_k^e(\mathbf{x}) + V^{nn}(\mathbf{x}); \tag{2.9}$$

the coupling operators $\hat{C}_{kk'}$ in the Equations (2.8) are determined by the electronic wavefunctions $\phi_i(\xi \mid \mathbf{x})$.

From the nuclear wavefunctions $\Psi_k(\mathbf{x})$ one is able to calculate the scattering matrix elements $S_{\alpha a, \beta b}$ which give the cross section by [5]

$$\sigma_{\alpha a, \beta b} \equiv \sigma(a \mid u_{\alpha a} \mid b) = (\pi \hbar^2 / P_{\alpha a}^2) \mid S_{\alpha a, \beta b} \mid^2 \tag{2.10}$$

where the letters α and β denote the reactant and product channels - the fragmentation of the whole system as given on the left-hand and right-hand sides of Equation (2.1), respectively - whereas a and b collect the reactant and product quantum numbers, respectively: $a \equiv (ij), b \equiv (lm)$; $P_{\alpha a}$ is the relative momentum of the reactants X_1 and X_2.

It should be mentioned that the relevant quantity σ, Equation (2.10), can be approximately calculated also by classical or quasiclassical methods.

(III) Kinetics of molecular ensembles. From the cross-section the rate constant is to be determined by summation and statistical averaging over distribution functions corresponding to the experimental situation [3, 4].

2.3. THE NUCLEAR MOTION PROBLEM: HEURISTIC CONSIDERATION

For the following we suppose that the process under consideration proceeds without change of the electronic state of the interacting system, i.e. the coupling operators $\hat{C}_{kk'}$ in Equation (2.8) are set equal to zero for $k \neq k'$ (electronically adiabatic approximation). Thus we have a uniquely defined potential energy function $U(\mathbf{x})$ in the nuclear dynamical problem (the index k indicating the electronic state can be dropped).

In treating the dynamics of the nuclei the difficulties from the dimensionality (i.e. the number $3N^n - 6$ of nuclear degrees of freedom) arise in a twofold way:

(a) in the large number of 'points' (nuclear configurations) $\mathbf{x}$ for which the electronic Schrödinger equation (2.6) is to be solved in order to obtain a sufficiently complete representation of the functions $E^e(\mathbf{x})$ and $U(\mathbf{x})$; this number is of the order of magnitude 10^F with $F = 3N^n - 6$;

(b) in the number of variables in the nuclear equations of motion.

Obviously we are in a dilemma: For systems with a few atoms, $N^n = 2$ or 3, the theoretical description of the (electronically adiabatic) dynamics can be largely mastered, the physical phenomena are well understood and we have many important contributions to the theoretical fundaments of reaction dynamics – but there is little practical interest in the results. In polyatomic systems with $N^n > 3$, on the other hand, chemists are much more interested, the theoretical treatment, however, is faced with rapidly growing methodical difficulties and computational expense. For the latter case drastic simplifications are necessary, and our aim is to achieve, by an appropriate formulation of the problem and systematic approximations, some 'minimal formulation' – so to speak, to throw away all ballast and to keep back only those features of the problem which are dynamically essential.

The starting point is the well-known empirical fact that molecular rearrangement processes are frequently local phenomena (fission of a bond, addition of an atom or a functional group at a specific position of a molecule and so on). From this localization assumption it follows that, for a fixed total energy E of the system,

– there should exist a preferred 'reaction path' and a progress variable ('reaction coordinate') along it, eventually a small number of such variables, which describe the gross course of the rearrangement whereas the remaining coordinates, because of the energy limitation, are restricted to relatively small variations;

– some part of the degress of freedom do not take part in the rearrangement, i.e. remain nearly unchanged in character and energy content.

Therefore one expects two possibilities for a reduced treatment:

(1) *static* (or *range*) *reduction*, i.e. reduction to that part of the nuclear configuration space which is energetically relevant, namely the reaction path and its near environmen

(2) *dynamic* (or *dimensional*) *reduction*, i.e. reduction to that subspace of the nuclear configuration space (corresponding to an 'active' part of the molecular system) which is dynamically relevant.

Clearly this idea is largely in accordance with the traditional chemical picture. The problem is how to make exhaustive use of this concept by incorporating it into a rigorous theory.

3. The Reaction Path Concept

The preparational step in realizing the idea of static and dynamic reduction is an analysis of the potential energy function, Equation (2.9),

$$U(\mathbf{q}) \equiv U(q^1, \ldots, q^F) \tag{3.1}$$

where F denotes the number of nuclear degrees of freedom ($F = 3N^n$ or $3N^n - 6$) and q^i are appropriate coordinates (mass-weighted cartesians, $m_\nu^{1/2} \mathbf{x}_\nu$ if ν numbers the nuclei, or some curvilinear coordinates etc.). Geometrically, the function U defines a hypersurface in an $(F + 1)$-dimensional space, the so-called potential energy surface (PES).

3.1. TOPOGRAPHY OF THE POTENTIAL ENERGY SURFACE

A stationary (critical) point q_0 of the PES is defined by the condition $\nabla U = 0$ at $q = q_0$ and is classified by the signs of the eigenvalues of the Hessian matrix $\mathbf{K}_0 \equiv (\partial^2 U/\partial q^i q^j)_{q=q_0}$.

If none of the eigenvalues is negative, the PES has a local minimum at $q = q_0$; this nuclear configuration corresponds to a stable or metastable molecular aggregate. If one of the eigenvalues is negative, the PES has a first-order saddle point at $q = q_0$ corresponding to a transition configuration of a molecular rearrangement. Other types of critical points play a minor role and will not be discussed here.

The *reaction path* (RP) for a molecular rearrangement is defined as the curve of minimal potential energy connecting two local minima (which may be located also in asymptotic regions of the PES where one or several coordinates become infinite, corresponding to fragmentation of the system) via a first-order saddle point. The so-defined RP (sometimes called also the 'minimum energy path' or the 'intrinsic reaction coordinate') as introduced by Fukui, Pechukas and others [6–9] is calculated as the steepest-descent path starting from the saddle point into both directions of the eigenvector belonging to the negative eigenvalue of the Hessian matrix.

The arc length s of this curve is properly to be identified with the *reaction coordinate*, and the parametric representation $q^i(s)$ of the RP satisfies the differential equation [8, 9]

$$dq^i/ds = -c^{-1/2}g^{ij}\partial U/\partial q^j \tag{3.2}$$

(in covariant form with the Einstein convention of summation over doubly appearing indices like j on the right-hand side) where

$$c = g^{ij}(\partial U/\partial q^i)(\partial U/\partial q^j)$$

is the norm of the gradient, $g^{ij} = \sum_k(\partial q^i/\partial x^k)(\partial q^j/\partial x^k)$ the metric tensor, x^k are cartesian coordinates (usually mass-weighted); the initial condition is $q_0^i = q^{\neq i}$ if the point $q^{\neq}$ denotes the saddle point configuration.

The fundamental character of this RP shows up in several special properties [7-9]: invariance against coordinate transformation, conservation of nuclear configuration symmetry along the RP between the stationary points, coincidence with a geodesic line.

The potential energy function along the RP,

$$U(q(s)) \equiv U_0(s), \tag{3.3}$$

is usually called the *potential profile* or reaction profile. It gives an idea about the energy requirements for the process under consideration to proceed, in particular the height of the activation barrier as the potential energy difference between the reactant nuclear configuration, $q = q^R$, and the transition configuration at the saddle point, $q = q^{\neq}$.

3.2. TRANSVERSE MODES

From the $3N^n$ nuclear degrees of freedom, six are so-called external modes (the overall translation and rotation of the nuclear frame), one is the motion along the RP, and the $3N^n - 7$ remaining are the so-called *transverse modes* which could be imagined as internal vibrations of the nuclear frame keeping a definite position on the RP. These transverse modes are obtained by means of a local normal mode analysis, pointwise along the RP. Thus we have to diagonalize the Hessian matrix (in mass-weighted cartesian coordinates x^j)

$$\mathbf{K}(s) = (\partial^2 U/\partial x^i \partial x^j)_{\mathbf{x}=\mathbf{a}(s)} \tag{3.4}$$

at a sequence of discrete positions on the RP given by vectors $\mathbf{x} = \mathbf{a}(s)$. Outside a stationary point, in general, the seven eigenvalues corresponding to the external and RP motions are not zero and the $3N^n - 7$ eigenvectors corresponding to the transverse modes are not orthogonal to the (known) eigenvectors of the external and RP motions [10]. This is connected with the fact that we have a nonvanishing linear term, $\nabla U \cdot (\mathbf{x} - \mathbf{a})$, in the Taylor expansion around the RP. These inconvenient properties can be removed by a projection procedure [10] leading to a new Hessian matrix

$$\mathbf{K}^P(s) = [\mathbf{1} - \mathbf{P}(s)]\mathbf{K}(s)[\mathbf{1} - \mathbf{P}(s)], \tag{3.5}$$

with the projecter $\mathbf{P}$ onto the directions of the seven eigenvectors of the external and RP motions. Now the diagonalization of $\mathbf{K}^P(s)$ gives, besides seven zero eigenvalues for these latter motions, the $3N^n - 7$ eigenvalues $\omega_k^2(s)$ for the transverse modes and the corresponding eigenvectors $\mathbf{L}_k(s)$ being now orthogonal to those of external and RP motions.

From the eigenvectors, mode coupling coefficients can be defined [10]:

$$B_{k,k'}(s) \equiv (\partial \mathbf{L}_k / \partial s) \cdot \mathbf{L}_{k'}(s), \tag{3.6}$$

If we number the reaction coordinate by $k = 1$ and the external mode coordinates by $k = 3N^n - 5, \ldots, 3N^n$, the expression

$$\kappa(s) \equiv \left[\sum_{k=2}^{3N^n-6} B_{k,1}(s) \right]^{1/2} \tag{3.7}$$

gives the RP curvature determined by the couplings of the transverse modes to the motion along the reaction coordinate.

Thus we get the following information from the local normal mode analysis at successive points on the RP:

(a) behaviour of the transverse mode frequencies (small or large amplitude/high or low frequency);

(b) behaviour of the coupling of all modes to each other.

It should be pointed out that this simple description of a molecular process in terms of reaction path and transverse modes has its limitations. Obviously it is not applicable in cases of strong RP curvature (because of the so-called multivaluedness problem) and in cases of RP bifurcations [11].

3.3. TWO EXAMPLES

3.3.1. The System O_2H. As a first example we will briefly discuss the system O_2H in which, besides the isomerization (intramolecular H transfer), bimolecular exchange from $H+O_2$ into $OH +O$ and unimolecular fragmentation of O_2H are adiabatically possible in the electronic ground state:

$$H(^2S) + O_2(^3{\textstyle\sum_g}) \;\overset{\rightarrow}{\leftarrow}\; O_2H(^2A') \;\overset{\rightarrow}{\leftarrow}\; OH(^2\Pi) + O(^3P). \tag{3.8}$$

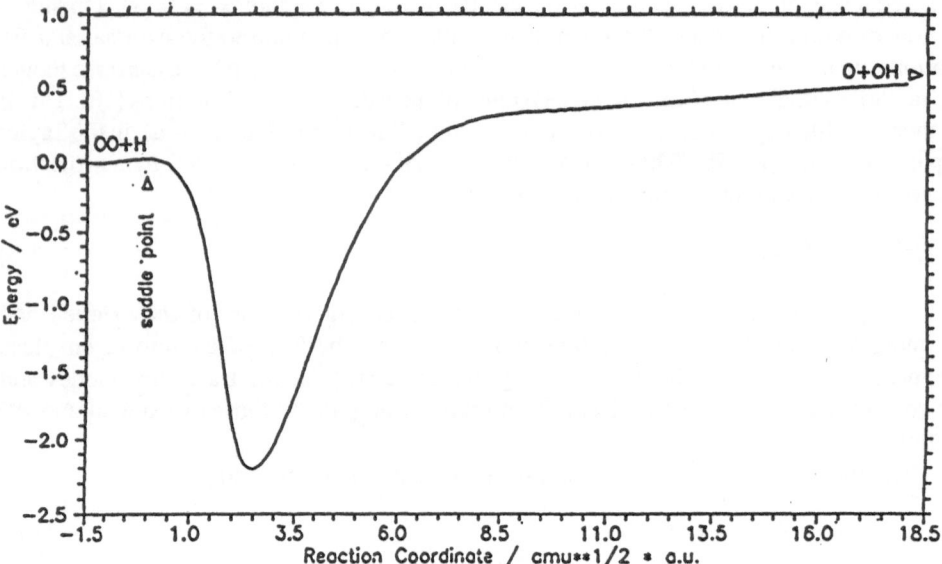

Fig. 2. Potential energy profile along the H+O₂ exchange reaction path (from [12]).

In Figure 2 the potential profile along the RP is seen to exhibit a pronounced minimum for the O₂H molecule; the process (3.8) is endoergic by 0.6 eV. Figure 3 shows the RP curvature with a pronounced double peak structure in the neighbourhood of the potential minimum position.

The transverse modes are determined by the potential energy in planes orthogonal to RP; contour line diagrams of this potential in the near vicinity of the RP for six values of the reaction coordinate s are plotted in Figure 4.

The most interesting feature is the change from a pair of stretching-type vibration (small amplitude) and bending-type vibration (large amplitude) in Figure 4(b) at the saddle point in the reactant channel to near-degeneracy in Figure 4(c), again to a pair of stretching and bending vibrations at the potential minimum in Figure 4(d) and subsequent deformation to a kidney shape whereby the bending mode transforms to a successively less hindered rotation (Figures 4(e), (f)).

The transverse mode frequencies in dependence on the reaction coordinate are shown in Figure 5; the transformation of the bending transverse mode into rotation is clearly seen as well as the near degeneracy where the splitting is so small that it cannot be resolved within the accuracy of the plot. Finally, we present in Figures 6 and 7 the coupling coefficients of the transverse modes to the RP and the coupling between the transverse modes, respectively.

A detailed analysis of this system is given in [12].

3.3.2. The System (HCH₃F)⁻. The rearrangement

$$H^-(^1S) + CH_3F(^1A_1) \rightarrow CH_4(^1A_1) + F^-(^1S), \tag{3.9}$$

proceeding adiabatically in the electronic ground state, can be considered as a prototype of bimolecular nucleophilic substitution (S_{N^2}). This system is already rather complicated and

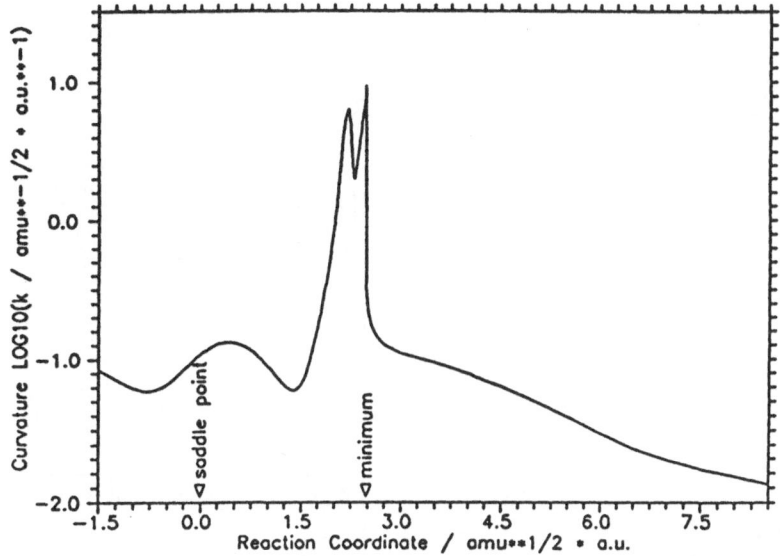

Fig. 3. Curvature of the H+O$_2$ exchange reaction path (from [12]).

we will restrict our discussion to some prominent features.

We start again with the potential profile as shown in Figure 8 with the well-known double-minimum structure, the small barrier of probably 5–15 kJ/mol (relative to the reactants) and the large exoergicity of 220–250 kJ/mol (compare, e.g., [13]). For the transverse modes we give in Figure 9 a correlation diagram between the reactants (R'), the pre-barrier complex (M') and the transition configuration (T).

Most of the frequencies do not change very much along this part of the RP which coincides at the bound complex M' with the asymmetric valence stretch vibration H$_{ax}$— CH$_3$—F, mode 1. The coupling coefficients [14] of the transverse modes to the RP, Figure 10, indicate that in the pre-barrier region only mode 2 is strongly coupled to the RP motion whereas in the vicinity of the transition configuration the coupling of modes 3 and 4 to the RP motion dominates.

4. Reaction Path Hamiltonians

As a next step, following the pioneering work of Miller, Handy, and Adams [10], one should formulate the theory in such a way that it is related to the topography of the potential energy function as closely as possible. To this end we derive advantage from information (a) of the local normal mode analysis thus achieving the static (or range) reduction mentioned in Section 2.3.

4.1. HAMILTONIAN FOR HARMONIC TRANSVERSE MODES

If all the transverse modes are of small-amplitude character at the relevant parts of the RP and if (for simplicity) the total angular momentum **J** of the system is taken as zero, normal

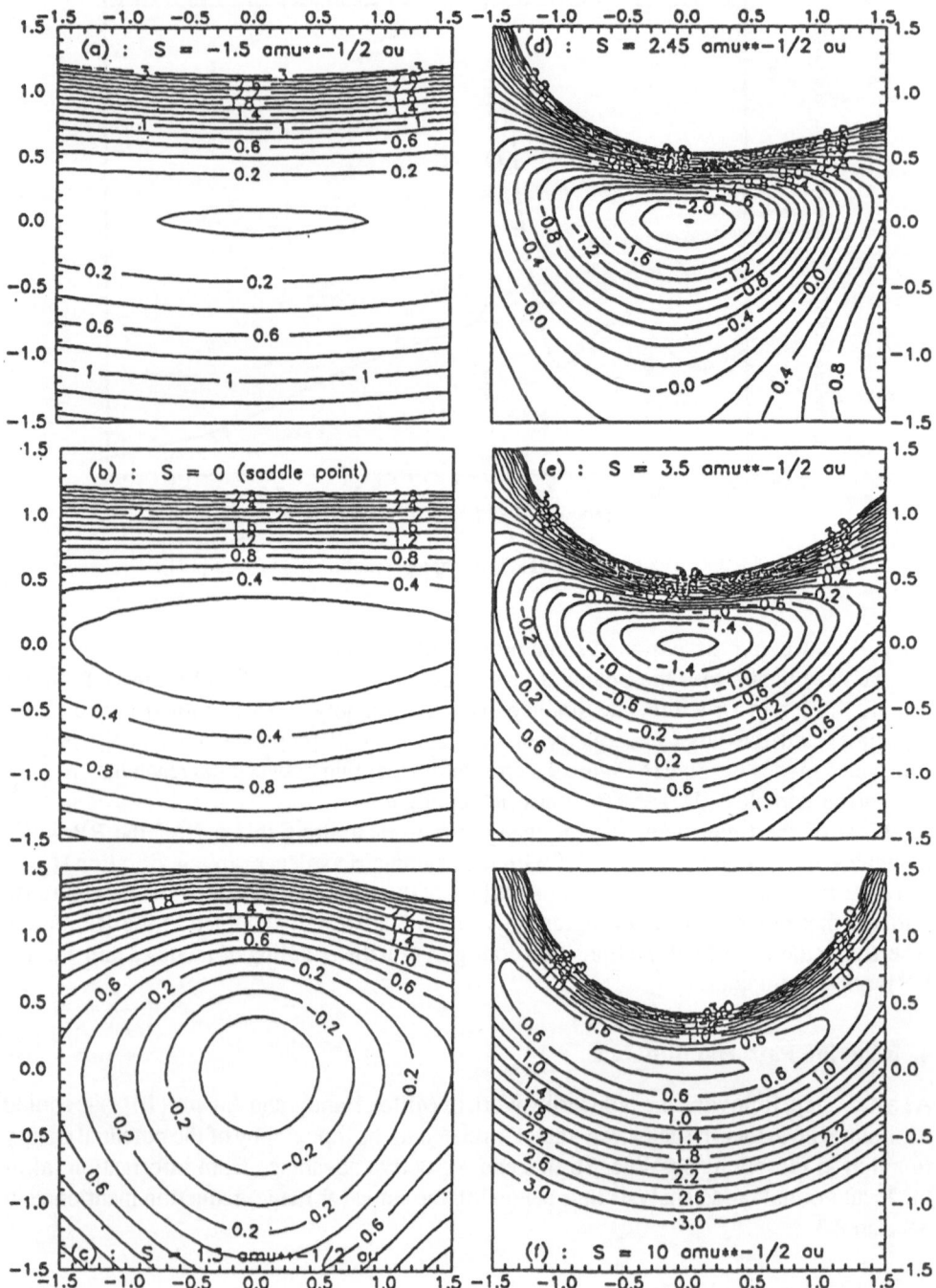

Fig. 4. Equipotential contour maps in planes orthogonal to the H+O₂ exchange reaction path (from [12]).

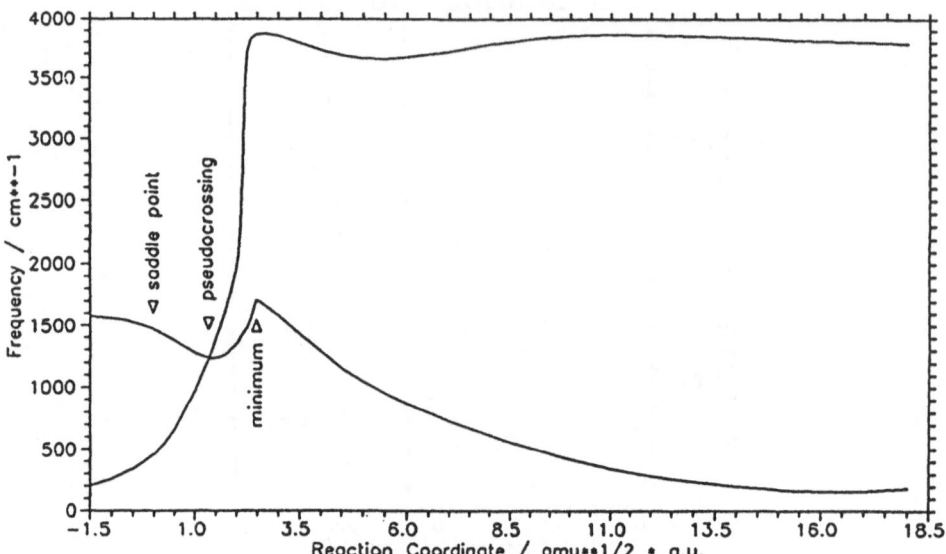

Fig. 5. Frequencies of the transverse modes along the $H+O_2$ exchange reaction path (from [12]).

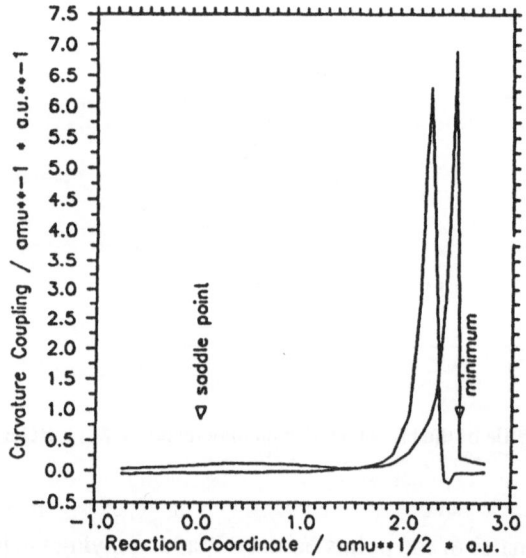

Fig. 6. Couplings of the transverse modes to the reaction path motion in the $H+O_2$ exchange reaction (from [12]).

coordinates Q^k can be used in the vicinity of the RP:

$$x^i = a^i(s) + \sum_{k=2}^{3N^n-6} L_{ik}(s)Q^k \qquad (i = 1, \dots, 3N^n - 6) \qquad (4.1)$$

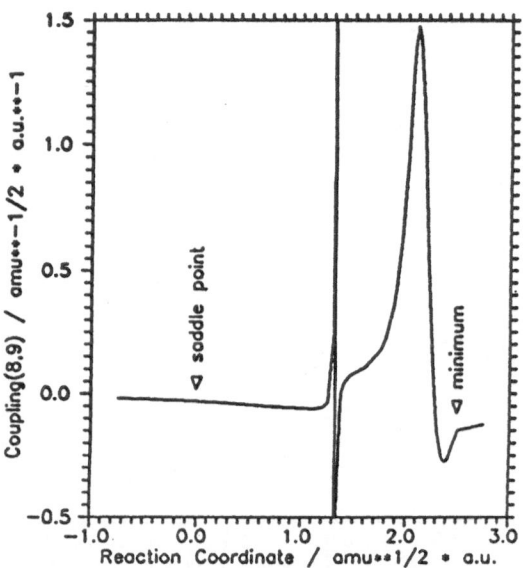

Fig. 7. Coupling of the transverse modes to one another along the H+O₂ exchange reaction path (from [12]).

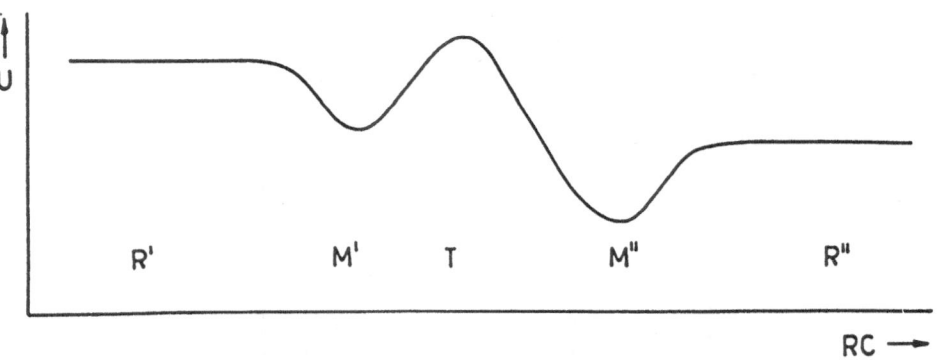

Fig. 8. Potential energy profile of the H⁻+CH₃F exchange reaction (S_{N2}). RC is a fictitious reaction coordinate (after [22]).

where a^i are the coordinates of the points on the RP, and applying the harmonic approximation to the transverse modes, the potential energy is given by:

$$U = U_0(s) + (1/2) \sum_{k=2}^{3N^n-6} \omega_k^2(s)(Q^k)^2. \tag{4.2}$$

By canonical transformation from the cartesian coordinates and momenta (x^i, p^i) to the reaction coordinate s and the transverse normal coordinates Q^k with the conjugated momenta p_s and P^k, respectively, the classical Hamiltonian for the internal motions of the

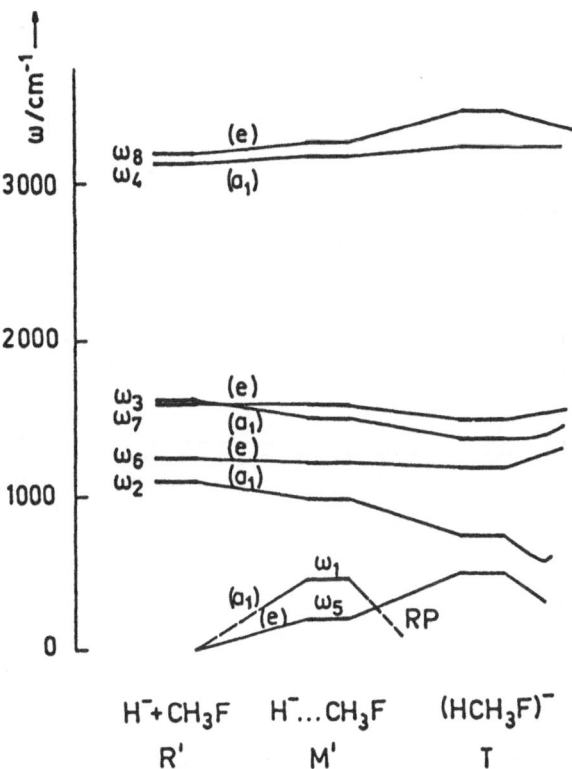

Fig. 9. Correlation diagram of the normal modes of C_3v-symmetric nuclear configurations RI, MI, and T in the H^-+CH_3F exchange reaction (from [22]). The a_1 modes represent the following motions: (1) asymmetric stretch $H_a x$—CH_3—F (at MI); (2) symmetric valence stretch $H_a x$—CH—F; (3) symmetric CH_3 ('umbrella') deformation; (4) totally symmetric C—H valence stretch in CH_3 (from [22]).

system becomes [10]

$$H(p_s, s, \{P^k, Q^k\}) = \tilde{p}_s^2/2\tilde{\mu}_s + U_0(s) + (1/2) \sum_{k=2}^{3N^n-6} [(P^k)^2 + \omega_k^2(s)(Q^k)^2] \quad (4.3)$$

(the so-called *reaction path Hamiltonian*) where the quantities

$$\tilde{p}_s \equiv p_s - \sum_{k \neq k'}^{3N^n-6} \sum_{=2}^{3N^n-6} Q^k P^{k'} B_{k,k'}(s), \quad (4.3a)$$

$$\tilde{\mu}_s \equiv [1 + \sum_{k=2}^{3N^n-6} Q^k B_{k,1}(s)]^2 \quad (4.3b)$$

represent a generalized momentum and a generalized effective mass, respectively. The first part of the Hamiltonian (4.3) clearly describes the motion along the RP (as usual, this is

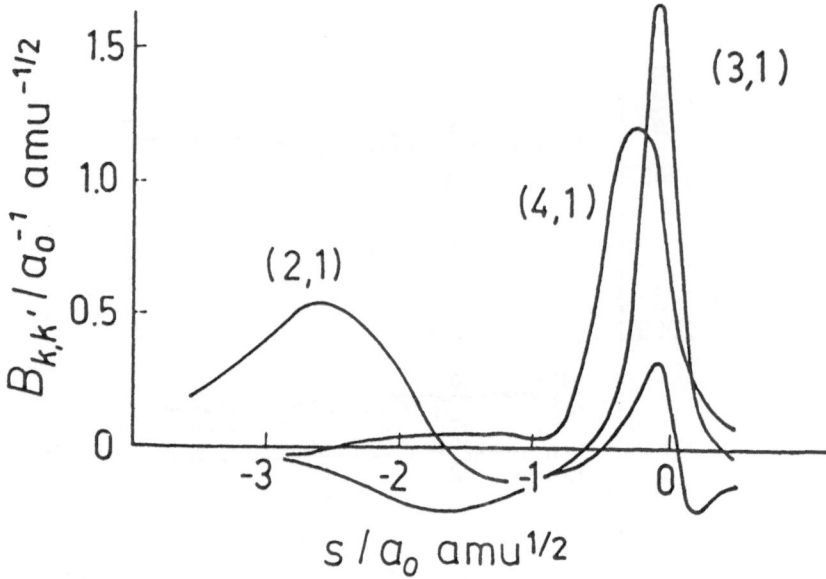

Fig. 10. Couplings of the transverse modes to the H^-+CH_3F exchange (S_{N2}) reaction path (from [14]).

mode 1) with the potential $U_0(s)$, whereas the second part is formally a normal vibration Hamiltonian for the transverse modes. Both parts are coupled, but the mode-mode coupling coefficients appear only in the kinetic energy of the first part. The corresponding quantum-mechanical Hamiltonian can be obtained by straightforward procedures [15].

This Hamiltonian, first derived by Miller, Handy, and Adams [10] and closely related to earlier work of Hofacker [16], Marcus [17], Basilevsky [8] and others, has interesting properties [10, 11] which will not be discussed here; it is an appropriate starting point for the formulation of simplified theories. Nevertheless it has some shortcomings which restrict the range of applicability; in particular the harmonic approximation to all transverse modes is of course not valid for bending vibrations transforming into rotations.

4.2. GENERAL FORMULATION

To remove restrictions as much as possible the following generalizations seem desirable [18]:

(i) formulation in curvilinear internal ('shape') coordinates since such coordinates can be chosen most closely adapted to the specific system, they may result in formally simpler expressions and lead eventually to smaller intermode couplings.

(ii) explicit account of large-amplitude modes (LAM) in addition to the RP as indicated by the local normal mode analysis discussed in Section 3.2.

The local normal mode analysis leads to a local splitting of the complete $3N^n$-

dimensional nuclear configuration space $\mathcal{E}$ (Nauts and Chapuisat [18]):

$$\mathcal{E} = \mathcal{E}_{RT} \oplus \mathcal{E}_{LAM} \oplus \mathcal{E}_{HM} \tag{4.4}$$

where $\mathcal{E}_{RT}$ is the six-dimensional subspace of the external degrees of freedom (overall translation and rotation of the system) and $\mathcal{E}_{LAM}$ is the $(n + 1)$-dimensional subspace of the RP motion plus n transverse modes which become of large amplitude somewhere along the RP. $\mathcal{E}_{HM}$ is the $(3N^n - n - 7)$-dimensional subspace of the remaining transverse degrees of freedom which are harmonic, i.e. of small amplitude, everywhere and can be determined by diagonalization of a projected Hessian matrix analogous to Equation (3.5) where now a local projector onto the space $\mathcal{E}_{RT} \oplus \mathcal{E}_{LAM}$ is to be used. In the example of O_2H discussed above, we have $n = 1$ (the bending mode with the lower of the two frequencies in Figure 5).

According to the different character of the modes in the subspaces, appropriate coordinates can be introduced: cartesian coordinates for the overall translation, eulerian angles for the overall rotation, the arc length of the RP as the reaction coordinate, some (in general curvilinear) coordinates v^i for the n large-amplitude modes ($j = 2, \ldots, n+1$) and normal coordinates Q^l ($l = n + 2, \ldots, 3N^n - 6$) for the harmonic modes.

The potential energy function has now the form

$$U = U_0(s) + U_{LAM}(v^1, \ldots, v^n, s) + (1/2) \sum_{l=n+2}^{3N^n-6} [\omega_l(v^1, \ldots, s)]^2 (Q^l)^2 \tag{4.5}$$

so that we get the general classical Hamiltonian for the internal motions of the system as

$${}^J H = {}^J T_{LAM} + U_0(s) + U_{LAM}(\mathbf{v}, s) + (1/2) \sum_{l=n+2}^{3N^n-6} [(P^l)^2 + \omega_l^2(\mathbf{v}, s) \ (Q^l)^2] \tag{4.6}$$

(for arbitrary total angular momentum J) where the kinetic energy part ${}^J T_{LAM}$ has a complicated form and will not be given here [18]. Again, the quantum-mechanical Hamiltonian can be obtained by standard procedures [15]. For the special case $n = O$, the Hamiltonian reduces to that by Miller, Handy and Adams, as it must be. In the version (4.6), the *generalized reaction path Hamiltonian*, the formalism has not been applied so far.

It should be noticed, finally, that a closely related generalization of the simple reaction path Hamiltonian, the so-called *reaction surface Hamiltonian*, has been elaborated and tested by Carrington and Miller [19]. Other generalizations being of more restricted nature exist which include, e.g., anharmonicity of transverse modes [20].

By a formulation of the theory as discussed here, based on the reaction path concept and retaining the full dimensionality of the problem, one can hope to achieve the most compact and system-adapted representation which should give the appropriate starting point for introducing well-founded simplifications.

4.3. SOME PRACTICAL PROBLEMS

At this stage of the discussion it seems worthwhile to point out that, now as ever, everything depends on whether one disposes of a correct potential energy function which governs the process considered. This still represents the main bottleneck in spite of the unquestionable progress of quantum chemical methodology. In particular, to get a correct potential energy profile U_0 with an accurate activation barrier is one of the most demanding tasks even for the simplest systems.

As to the determination of the other ingredients of a reaction path Hamiltonian, appropriate techniques (search and optimization procedures, analytical derivative methods incorporated into quantum chemical programs) become more and more available so that the reaction path, transverse mode frequencies and eigenvectors, and from them the coupling coefficients are expected to be soon routinely accessible (compare [21]) – hopefully with sufficient accuracy.

5. Dimensional Reduction of the Nuclear Dynamics

The final step in following the strategy sketched in Section 2.3. is to use information (b) from the local normal mode analysis of Section 3.2., i.e. to perform the dynamical (or dimensional) reduction (compare [22]).

5.1. PARTITIONING OF THE NUCLEAR DEGREES OF FREEDOM

In the model hierarchy we are now on the level of internal nuclear dynamics. According to the scheme in Figure 1 we divide the set S of nuclear degrees of freedom into two subsystems: S_1, the 'active subsystem' (sometimes it will simply be called the 'system'), containing the reaction path motion plus those transverse modes which are strongly coupled to it (altogether f degrees of freedom where f is hopefully considerably smaller than the total number F of degrees of freedom in S), and S_2, the 'inactive subsystem' (sometimes called the 'bath'), consisting of the remaining transverse modes which are weakly coupled to the modes of S_1 (altogether $F - f$ degrees of freedom).

This grouping of the modes can be accomplished, along the lines of our arguments in the preceding sections, by an analysis of the classical properties of the modes, in particular the coupling coefficients $B_{k,k'}(s)$, or by corresponding quantum-mechanical decoupling procedures. More simple, consideration of characteristic times for the energy exchange between the modes or even intuitive structural arguments may be applied. Note that no reference is made here to whether a transverse mode is of harmonic or large-amplitude type.

For illustration, we return briefly to our two examples discussed in Section 3.3. In the system HO_2, for the $H+O_2$ recombination (or the corresponding HO_2 fragmentation) it should be a good approximation to put the RP motion (eventually plus the transverse stretch mode) into the active subsystem. In the pre-barrier region which turns out to be most important for the S_N2 rearrangement $H^- + CH_3F$ (vide infra) the active subsystem should be composed of the RP motion plus mode 2, at least.

5.2. DYNAMICAL TREATMENT

If we now suppose that the process under consideration is decisively determined by the dynamics of the active subsystem S_1 we are led to an obvious simplified procedure: to treat the dynamics of S_1 explicitly as accurate as possible, taking into account the influence of S_2 in an approximate way.

Various possibilities to do this are presently under discussion and have been tested in same cases; they are almost exclusively based on the Miller-Handy-Adams formulation. We give here a brief summary without details.

(1) *Vibrationally adiabatic approximation*:

If the modes of S_2 are much faster than those of S_1, a vibrationally adiabatic separation between the two subsystems may be a good approximation. The quantum numbers n_k of the bath modes ($k = f+1, \ldots, F$) are kept constant; the corresponding energy of the bath modes in the harmonic approximation,

$$U_{\mathbf{n}} = \sum_{k=f}^{F} n_k + 1/2)\hbar\omega_k(s, \ldots), \tag{5.1}$$

added to the potential energy function of the active modes, gives an effective potential governing the dynamics of these modes [10].

A special case of such an approach is to replace the dynamical description of the active subsystem by a statistical treatment, in particular for the case $f = 1$, i.e. the active subsystem consists of the reaction coordinate only. Thus the procedure reduces to a version of the Generalized Transition State Theory [23] or the statistical Adiabatic Channel Model [24]. The primary result from this kind of theory is a rate constant according to the statistical assumptions (microcanonical ensemble, e.g.).

(2) *Perturbation-theoretical correction of the S_1 dynamics*:

After a separate treatment of the dynamics of the active modes, the influence of the inactive modes can be approximately taken into account by an appropriate perturbation-theoretical procedure [25].

(3) *Stochastic treatment of the S_1 dynamics*:

A tempting idea, particulary in cases with a large number of inactive degrees of freedom, is to take the influence of the bath into account by friction and random force terms acting on the active degrees of freedom; the dynamics of the latter is then described by generalized Langevin equations [11, 26].

5.3. APPLICATIONS

It is not the aim of the present paper to give a complete account of all the experience gained in dimension-reduced treatments of molecular processes; a compilation of numerous applications related to this concept is given elsewhere [22]. Most of the studies so far are restricted to the RP dynamics only ($f = 1$) or corresponding statistical treatments. Only a few attempts have been made to consider cases with $f > 1$; we mention here the studies of the $H+H_2$ exchange process ($f = 2$) by Schwartz and Miller [25], the malonaldehyde intramolecular H transfer ($f = 2$) by Carrington and Miller [27] and the S_N2 processes $X^- + CH_3F$ with X=H or halogen ($f = 2$) by Basilevsky and Ryaboy [28].

For a particular case in the latter investigation, the $_SN2$ rearrangement $H^- + CH_3F$ as considered in Section 3.3.2., a comparison is possible with a corresponding statistical adiabatic treatment [29]. Both studies use the same quantum-chemically calculated input data (to give some characteristics of the potential profile: 35.1 kJ/mol for the well depth $\Delta R'M'$ and 4.6 kJ/mol for the barrier height $\Delta TR'$ relative to the reactants).

Supposing a three-step mechanism according to Brauman *et al.* [30]:

(1) $R' \rightarrow M'$, formation of the complex M', (2) $M' \rightarrow M''$, unimolecular conversion into complex M'' via the transition configuration T, (3) $M'' \rightarrow R''$, fragmentation of the complex M'' into the products; assuming steady-state concentrations of M' and M'' and taking into account the strong exothermicity of the reaction (see Figure 8), the overall rate constant is approximately given by

$$k \approx k_1 k_2 (k_{-1} + k_2)^{-1} \tag{5.2}$$

(i.e. the ion-molecule capture rate constant times the forward-backward branching factor). The three relevant rate constants k_1, k_{-1} and k_2 are calculated on the basis of statistical models (for the details see [29]) leading to the thermal rate constant $k = 1.6 \times 10^{-11}$ cm^3 s^{-1} at $T = 300$ K. The dimension-reduced quantum-mechanical treatment of BASILEVSKY and RYABOY [28] gives $k = 1.1 \times 10^{-11}$ cm^3 s^{-1} at this temperature. In consideration of the substantial differences of the two approaches, this close agreement of the results is unexpected and certainly fortuitous. Furthermore, both theoretical values are in accordance with an experimental rate constant of 1.5×10^{-11} cm^3 s^{-1} [31]. In spite of all necessary reservation in drawing conclusions from these results they should at least give some hope that the theoretical approaches (including the potential energy data as their common starting point) are of some significance.

6. Conclusions and Prospect

The theoretical treatment of polyatomic reactions requires drastic approximations in order to cut the computational effort and to enlarge the range of applicability of theoretical methods. In the development of conceptually transparent and computationally feasible theoretical models a certainly prospective way is to restrict the treatment by static and dynamic reduction of the nuclear part of the dynamical problem, based on the reactionpath description and decoupling considerations as described in this paper.

Generalized reaction path Hamiltonians have been already formulated, but practicable versions and model approaches (including semi-empirical schemes) are still to be worked out and tested. Hopefully, the input data of such a theory (electronic energy and energy derivative quantities) will be more and more routinely accessible (an essential role play analytical derivative techniques); in many cases, however, the requirements on accuracy, e.g. for activation barriers, are extremely difficult to fulfil.

Acknowledgements

The author thanks Dr. A. Merkel and Dr. F. Schneider (both Berlin) for their contributions to the subject over several years. The cooperation with Prof. M.V. Basilevsky and Dr. V.M. Ryaboy (both Moscow) stimulated the progress very much. Last but not least we profited from the common work with Prof. X. Chapuisat (Orsay) in the conceptual part as well as in the O_2H case study reported here.

References

[1] H. Eyring: *J. Chem. Phys.* **3** (1935) 107.
[2] H. Pelzer and E. Wigner: *Z. phys. Chem.* **B15** (1932) 445.
    E. Wigner: *Trans. Faraday Soc.* **34** (1938) 29.

[3] R. D. Levine and R. B. Bernstein: *Molecular Reaction Dynamics and Chemical Reactivity*, Oxford University Press, New York and Oxford, 1987.

[4] E. E. Nikitin and L. Zülicke: *Selected Topics of the Theory of Chemical Elementary Processes*, (Lecture Notes in Chemistry, Ed. G. Berther *et al.*, Vol. 8). Springer Verlag, Berlin-Heidelberg-New York, 1978.

[5] L. D. Landau and E. M. Lifshitz: *Quantum Mechanics. Non-Relativistic Theory*, 2nd Ed., Addison-Wesley, Reading, Mass. 1965.

[6] K. Fukui: *J. Phys. Chem.* **74** (1970) 4161.

[7] P. Pechukas: *J. Chem. Phys.* **64** (1976) 1516.

[8] M. V. Basilevsky: *Chem. Phys.* **24** (1977) 81.

[9] A. Nauts and X. Chapuisat: *Chem. Phys. Lett.* **85** (1982) 212.

[10] W. H. Miller, N. C. Handy, and J. E. Adams: *J. Chem. Phys.* **72** (1980) 99.

[11] W. H. Miller: *J. Phys. Chem.* **87** (1983) 3811.

[12] F. Schneider, L. Zülicke, and X. Chapuisat: *Mol. Phys.* **71** (1990) 17.

[13] Z. Havlas, A. Merkel, T. Kalcher, R. Janoschek, and R. Zahradnik: *Chem. Phys.* **127** (1988) 53.
R. Vetter and L. Zülicke: *J. Mol. Struct. (Theochem.)* **170** (1988) 85.

[14] V. M. Ryaboy: *Chem. Phys. Lett.* **159** (1989) 371.
A. Merkel: unpublished results.

[15] B. Podolsky: *Phys.Rev.* **32** (1928) 812.

[16] G. L. Hofacker: *Z. Naturforsch.* **A18** (1963) 607.

[17] R. Marcus: *J. Chem. Phys.* **45** (1966) 4493, 4500;
ibid. **49** (1968) 2610.

[18] A. Nauts and X. Chapuisat: *Chem. Phys.* **76** (1983) 349.

[19] T. Carrington and W. H. Miller: *J. Chem. Phys.* **81** (1984) 3942.

[20] D. G. Truhlar, F. B. Brown, R. Steckler, and A. D. Isaacson: in *The Theory of Chemical Reaction Dynamics* (Ed. D. C. Clary), D. Reidel, Dordrecht, 1986.

[21] B.C. Garrett, M. J. Redmon, R. Steckler, D. G. Truhlar, K. K. Baldridge, D. Bartol, M. N. Schmidt, and M. S. Gordon: *J. Phys. Chem.* **92** (1988) 1476.
M. Page and J. W. McIver: *J. Chem. Phys.* **88** (1988) 922.

[22] L. Zülicke AND A. Merkel: *Int. J. Quantum Chem.* **38** (1990) 191.

[23] B. C. Garrett and D. G. Truhlar: *J. Phys. Chem.* **83** (1979) 1052, 1078.
D. G. Truhlar, A. D. Isaacson and B. C. Garrett: in *Theory of Chemical Reaction Dynamics*, (Ed. M. Baer), CRC Press, Boca Raton, 1985.

[24] M. Quack and J. Troe: *Ber. Bunsenges. Phys. Chem.* **73** (1974) 240.

[25] W. H. Miller and S. D. Schwartz: *J. Chem. Phys.* **77** (1982) 2378.
S. D. Schwartz and W. H. Miller: *J. Chem. Phys.* **79** (1983) 3759.

[26] W. H. Miller: in *Potential Energy Surfaces and Dynamics Calculations*, (Ed. D. G. Truhlar), Plenum, New York, 1981.

[27] T. Carrington and W. H. Miller: *J. Chem. Phys.* **84** (1986) 4364.

[28] M. V. Basilevsky and V. M. Ryaboy: *Chem. Phys. Lett.* **129** (1986) 71.
V. M. Ryaboy and A. Merkel: *Fiz. Khim. (Moscow)*, in press, and unpublished results.

[29] Z. Havlas, A. Merkel, and R. Zahradnik: *J. Am. Chem. Soc.* **110** (1988) 8355.
A. Merkel: Thesis (A), Academy of Sciences of the GDR, Berlin, 1986.

[30] W. N. Olmstead and J. I. Brauman: *J. Am. Chem. Soc.* **99** (1977) 4219.

[31] K. Tanaka, G. I. Mackay, J. D. Payzant, and D. K. Bohme: *Can. J. Chem.* **54** (1976) 1643.

AN APPLICATION OF THEORETICAL MODELS TO UNDERSTAND CHEMICAL REACTIONS: THE ELECTROPHILIC SUBSTITUTION IN FURAN

J. RAUL ALVAREZ-IDABOY, LUIS A. MONTERO and RICARDO MARTÍNEZ
Facultad de Química-IMRE
Universidad de La Habana
Havana 10400
Cuba

ABSTRACT. Theoretical molecular modelling of furan ring substitution by electrophilic agents is reviewed. Both semiempirical and *ab initio* SCF MO calculations are extensively performed, including transition state and σ complex structures to propose mechanisms which can explain the unusual behaviour of such systems. A case study referring to alkenyl furan substitution is described where models are particularly successful in explaining experimental facts.

1. The mechanism

The main purpose of this paper is to illustrate the ability of theoretical methods to help chemists understand processes which are decided on the molecular level, where no instruments are available to measure them directly. In our experience, furan chemistry is full of unexpected behaviours.

Generalizations from the chemistry of aromatic hydrocarbons have frequently distorted the understanding of many phenomena, which often need their own explanation. That is why we chose the electrophilic substitution in furan ring as the subject of our theoretical research. This paper addresses some of our results in a comprehensive way, and is intended not only for theoreticians, but also for experimentalists who may find many of our results surprising – like furan chemistry itself. Likewise, the methodology of our research can be applied to other cases of interest.

1.1. THE REACTION ITSELF

From the phenomenological point of view, electrophilic substitutions in aromatic and other conjugated compounds are very well understood by organic chemists. However, there are very few papers which try to model these reactions from a theoretical molecular point of view [1], especially when more efficient procedures are available. These reactions consist of the attack of an electrophilic species to a conjugated π bond. It conducts to a final cleavage of σ bonds between the involved atoms of the substrate to substitute hydrogen – it could be another chemical group – by the electrophile active group. The usual steps include the ephemeral existence of a so-called σ *complex*, which temporarily binds together

L. A. Montero and Y. G. Smeyers (eds.), Trends in Applied Theoretical Chemistry, 71–98, 1992.

I

both the electrophilic group and the existing substituent in the same atom, and temporarily destroys conjugation in such site.

The corresponding energy evolution path is characterized by a previous maximum which is known as the *activation energy* of the σ complex step. It implies a transient molecular geometry meeting both reacting molecules, mutually perturbed in their chemical structures. This intermediate state is called a *transition state* and could be also understood as an energetic minimum in all but the reaction coordinate. Transition states are represented as saddle points in the corresponding hypersurface, that is, the n dimensional surface which gives the value of energy as a function of n independent geometrical variables.

The above mentioned mechanisms have been extensively studied and had been deduced from isotopic effect studies [2, 3] and NMR spectroscopy in non nucleophilic solvents at low temperatures [4, 5]. These studies corroborate the actual existence of the σ complex.

On the other hand, gas phase studies, both experimental [6–8] and theoretical [6–9], have confirmed that reactions where electrophiles are very strong, i.e. molecular cations, occur without a significant activation energy, and the resulting σ complex is more stable than both reactants and final products.

Therefore, decreasing strength in the polarizing capacity of the electrophile must change the reaction coordinate energy profile from a monotonic decrease in the case of cations, to the typical behaviour in solution [10] or solvated electrophiles as described above. This extreme case is the most usual, and must involve such a transition state, occurring before and being very similar to the σ complex.

Some of our papers [11–13] have shown the electrophilic substitution in five membered heterocycles systems as Furan (I) from the point of view of a very extended criterion based in the *Hammond's postulate* [14], which states that *the σ complex is the intermediate species determining the site and substrate reactivity when the electrophile strength is sufficiently small for allowing the formation of transition states with a similar structure to that of the σ complex.*

In our previously published articles and in parts of this present paper, the gas phase reaction has been theoretically modelled assuming that the solvent effect is the same, at least qualitatively, for all studied reactions. This must be true when solvation energies of σ complexes in both α and β ring I sites are expected to be similar. This can be deduced from the regularities in the experimental site preferences. However, it is desirable to build a theoretical molecular model where different reaction steps can be compared from several points of view, as energy preferences, geometrical features and the amount of *transferred charge* to confirm such criteria.

By using the semiempirical AM1 [15] method, which is contained in the current AMPAC program [16], many possibilities for following reaction paths and locating transition states

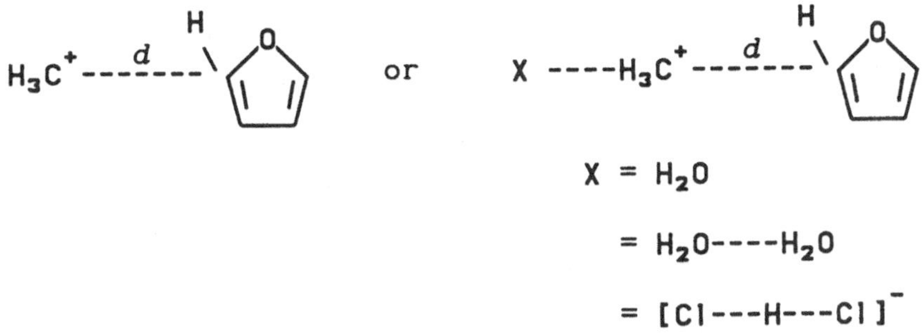

$$X = H_2O$$
$$= H_2O----H_2O$$
$$= [Cl---H---Cl]^-$$

II

are available. The tools for this theoretical modelling are described in the Appendix.

The selected reaction coordinate d is shown in II. All supermolecular conformations have been obtained by departing from the optimized geometries of the points near to the maximum energy in the reaction path.

For our purposes, critical points of its methylation reaction path with different selected agents as CH_3^+, $CH_3^+-H_2O$, $CH_3^+-2H_2O$, and $CH_3Cl-HCl$ have been studied. Of these electrophiles, the first three are very strong and the last one is relatively weak, from a qualitative point of view. In our model, water and HCl can be considered as *leaving groups* in the substitution reaction.

In the transition state neighbourhood, the distance between the electrophile carbon atom and the water oxygen in $CH_3^+-H_2O$ and $CH_3^+-2H_2$ – or chlorine in $CH_3Cl-HCl$ – is also varied, in order to look for the best minimum in respect to this coordinate in the transition state. It is taken as the input geometry for the gradient norm minimization method s(see Appendix).

The most critical points in the research path to be optimized are: i) the *isolated reactants*; ii) the *prereactive complex*, which appears corresponding to a slight energy minimum that must be due predominantly to electrostatic association in certain reaction paths (Figs. 1a and 2a); iii) the *transition state*, which is modelled as the absolute maximum in the reaction path which appears immediately before σ complex formation. It is coincident with the point in which the methyl group is partially linked to both the aromatic ring site and H_2O – or $[Cl-HCl]^-$ complex (Figs. 1b and 2b); iv) the σ *complex*, which is modelled as the minimum in the reaction path that corresponds to the transitory cation formed by the addition of the methyl group to Furan (Figs. 1c and 2c).

In the cases where methylation occurs by $CH_3^+-H_2O$, some equivalent critical points are found after the substitution which correspond to proton elimination, i.e. the final steps in the reaction.

Table I shows the energetic details of the calculated molecular and supermolecular structures for each one of the studied electrophilic agents. Figs. 3a to 3d show the energy profiles of the reaction paths.

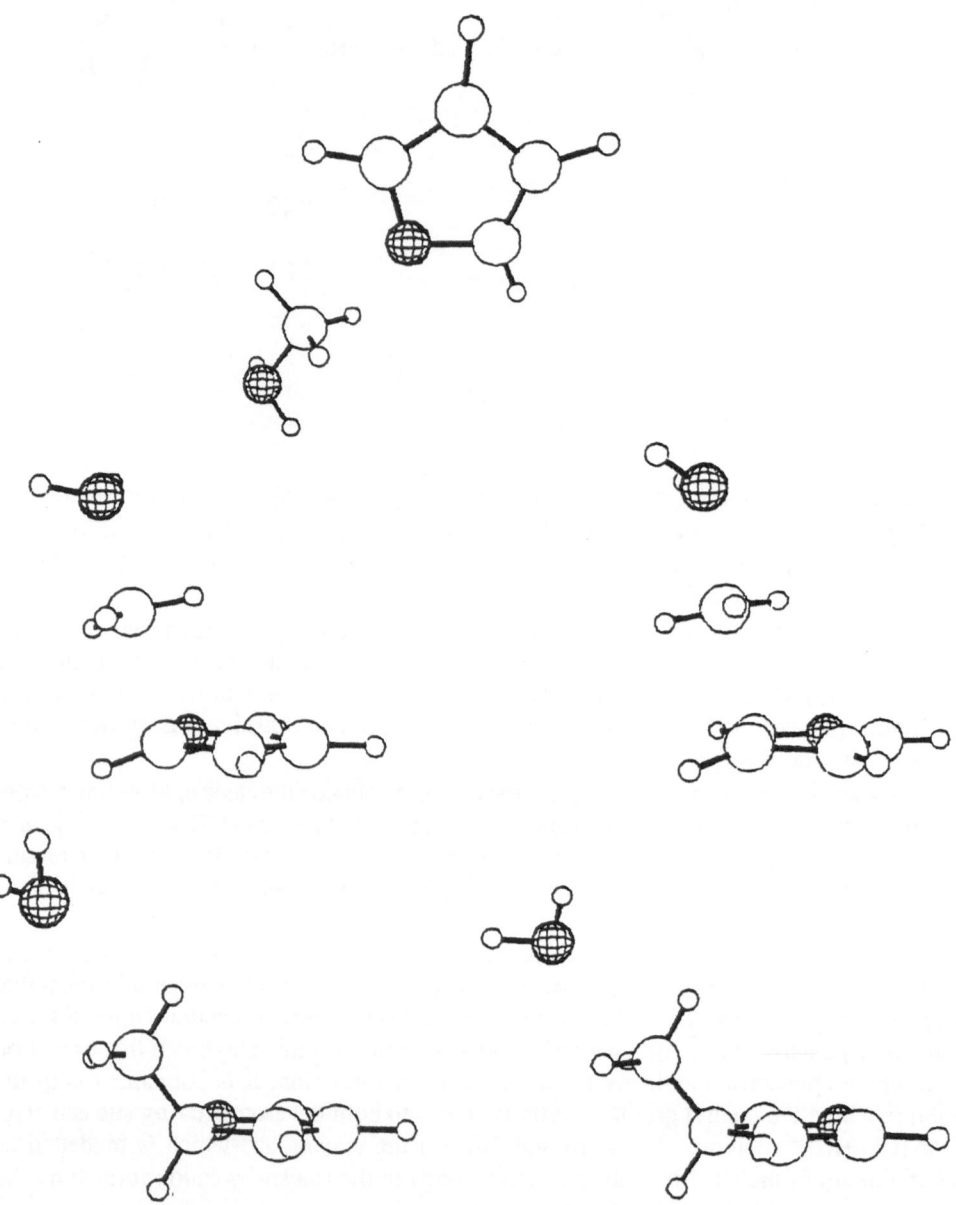

Fig. 1. Molecular graphics of fully AM1 optimized structures of three steps in the reaction path of $CH_3^+ - H_2O$ electrophilic attack to Furan. a) Prereactive complex; b) Transition state by the C_α site (up) and C_β site (down) (distances between ring carbons and CH_3^+ are larger than the index of covalent bond distances to not draw a bond stick); c) σ Complex by C_α site (up) and C_β site (down).

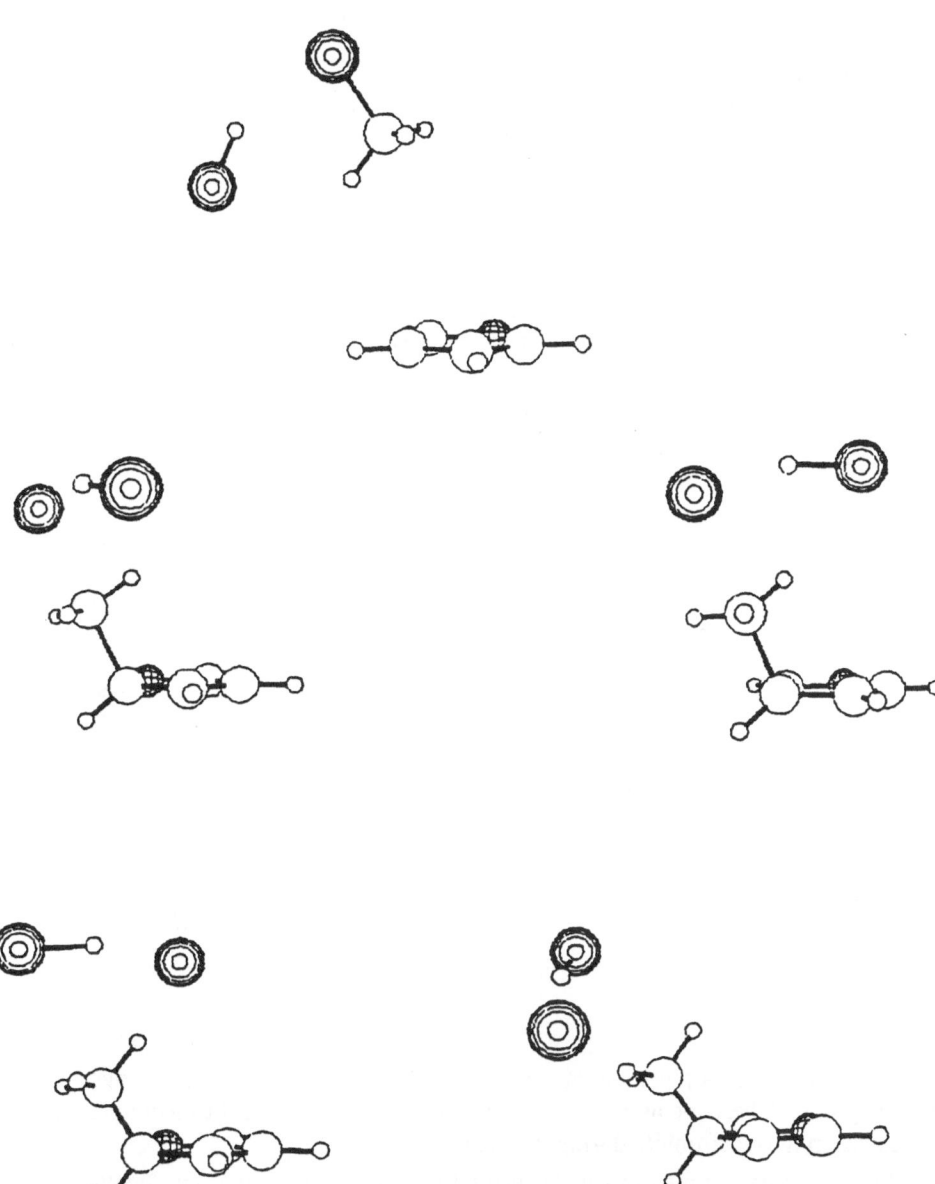

Fig. 2. Molecular graphics of fully AM1 optimized structures of three steps in the reaction path of $CH_3^+ Cl-HCl$ electrophilic attack to Furan. a) Prereactive complex; b) Transition state by the C_α site (up) and C_β site (down) (distances between ring carbons and CH_3^+ are less than the index of covalent bond distances to draw a bond stick); c) σ Complex by C_α site (up) and C_β site (down).

TABLE I

Calculated AM1 heats of formation for the reaction species in kJ/mole.

Reactants	Species		AM1 ΔH_f	ΔH_{rel}
$C_4H_4O + CH_3$	Isolated reactants		1068.2	0.0
	C_α	Sigma complex	701.4	−366.8
	C_β	Sigma complex	739.1	−329.1
$C_4H_4O + CH_3 \cdot H_2O$	Isolated reactants		591.0	0.0
	Pre-reactive complex		568.9	−22.1
	C_α	Transition state	612.5	21.5
		Sigma complex	430.0	−161.0
	C_β	Transition state	616.5	25.5
		Sigma complex	468.8	−122.2
	Proton elimination			
	C_α	Transition state	521.7	−69.3
		Postreac. complex	495.4	−97.6
	C_β	Transition state	524.4	−66.6
		Postreac. complex	521.7	−69.3
$C_4H_4O + CH_3 \cdot 2H_2O$	Isolated reactants		254.8	0.0
	Pre-reactive complex		236.9	−17.9
	C_α	Transition state	307.9	53.1
		Sigma complex	157.3	−97.5
	C_β	Transition state	311.0	56.2
		Sigma complex	196.1	−58.7
$C_4H_4O + CH_3Cl \cdot HCl$	Isolated reactants		−171.1	0.0
	Pre-reactive complex		−176.9	−5.8
	C_α	Transition state	71.1	242.2
		Sigma complex	10.7	181.8
	C_β	Transition state	89.7	260.8
		Sigma complex	54.2	225.3

The selected electrophile is always CH_3^+, alone or accompanied by the above mentioned leaving groups. These are understood to be performing the role of the most influent solvent molecules in the electrophile during the actual reaction.

The lowest unoccupied molecular orbital (LUMO) eigenvalues can be considered as related to the strength of each of the composed electrophilic species, i.e. including their respective leaving groups. AM1 method gives −9.38 ev for CH_3^+, −5.99 ev for $CH_3^+ - H_2O$, −5.05 ev for $CH_3^+ - 2H_2O$, and 1.20 ev for $CH_3Cl - HCl$. These values clearly illustrate the above mentioned qualitative order of electrophilicity.

The following is our comment concerning the different reaction steps:

Prereactive complex: The first relevant feature in Figs. 3b to 3d is the left hand minima, which represent the above mentioned *prereactive complex*. Figs. 1a and 2a show a molecular graphic of the optimized structure corresponding to a typical and representative case of such a stage in the reaction path. The graphics correspond to the methylation by the monohydrated electrophile and the hydrochlorinated methyl chloride. This supermolecule could also be understood as a π complex in classical aromatic substitutions. However, in these and the other cases studied, it appears as an association of the electrophile and its leaving group with the heterocyclic oxygen. It must be driven by electrostatic attractions, while the AM1 calculated gross atomic charges give a significant negative value to the heteroatom.

Transition state and σ complex: Environmental effects in these reactions can be deduced from the influence of the electrophile leaving group on the σ complex structure. Bond distances results between the methyl carbon atom and the leaving group appear around 2.6 Å in the cationic reactions (where water is the leaving group) and 3.15 Å in the case of the neutral electrophile. These values are in the order of van der Waals interactions, and could be taken as an evidence of certain degree of σ complex solvation. However, no differences are observed in the theoretical geometry of all of them among different electrophiles, including the direct methylation. This result can be taken as an indication of stability in the σ complex geometry in respect to different media – and *enables further calculations of such σ complexes without having to take into account the environmental effects.*

Regarding differences between C_β type σ complex energies in respect to those in C_α, the values of 37.7 kJ/mole for non hydrated cation, 38.8 kJ/mole for both the mono- and bihydrated cation, and 37.9 kJ/mole for hydrochlorinated methyl chloride – see Table I – are quite the same.

The results of AM1 calculations for the hydrochlorinated electrophile reproduce Hammond's postulate assumptions about weak electrophiles, that must lead to delayed transition states, which present very strong interactions with the substrate and determine the future site of substitution between C_α and C_β positions. Therefore, transition states formed by weak electrophiles are very similar to the σ complex, and reproduce the expected site selectivity.

On the other hand, when electrophiles are sufficiently strong, like all other calculated cases, transition states are attained early in the reaction path during the approach between reacting molecules, which implies more weak interactions in the supermolecule and a consequent decrease in selectivity. It is determined by the higher stability of the resulting σ complex during a dynamic equilibrium.

Table I and Fig. 3a show the expected lack of activation energy in the extreme case where isolated cation is the methylating agent. $CH_3^+ - H_2O$, and $CH_3^+ - 2H_2O$ systems are in the intermediate relative position between the strongest and the weakest electrophiles. They are very reactive and poorly selective as usual cations, but show an appreciable activation energy. In both cases, there is very little difference between C_α and C_β sites. However, slight differences favor a larger relative probability for surpassing the activation barrier to the C_α site, which can be calculated according to a Boltzmann distribution of 0.815 (81.5%) at room temperature (298 K) in the case of the monohydrated methylation.

Both transition states of such hydrated species are characterized by an electrophile-substrate distance of about 2.2 Å. These structures represent classical SN2 transition states. It must be pointed out that the reaction profiles here are closer to the gas phase model, as evidenced by the calculated larger stability of σ complexes with respect to the reactants.

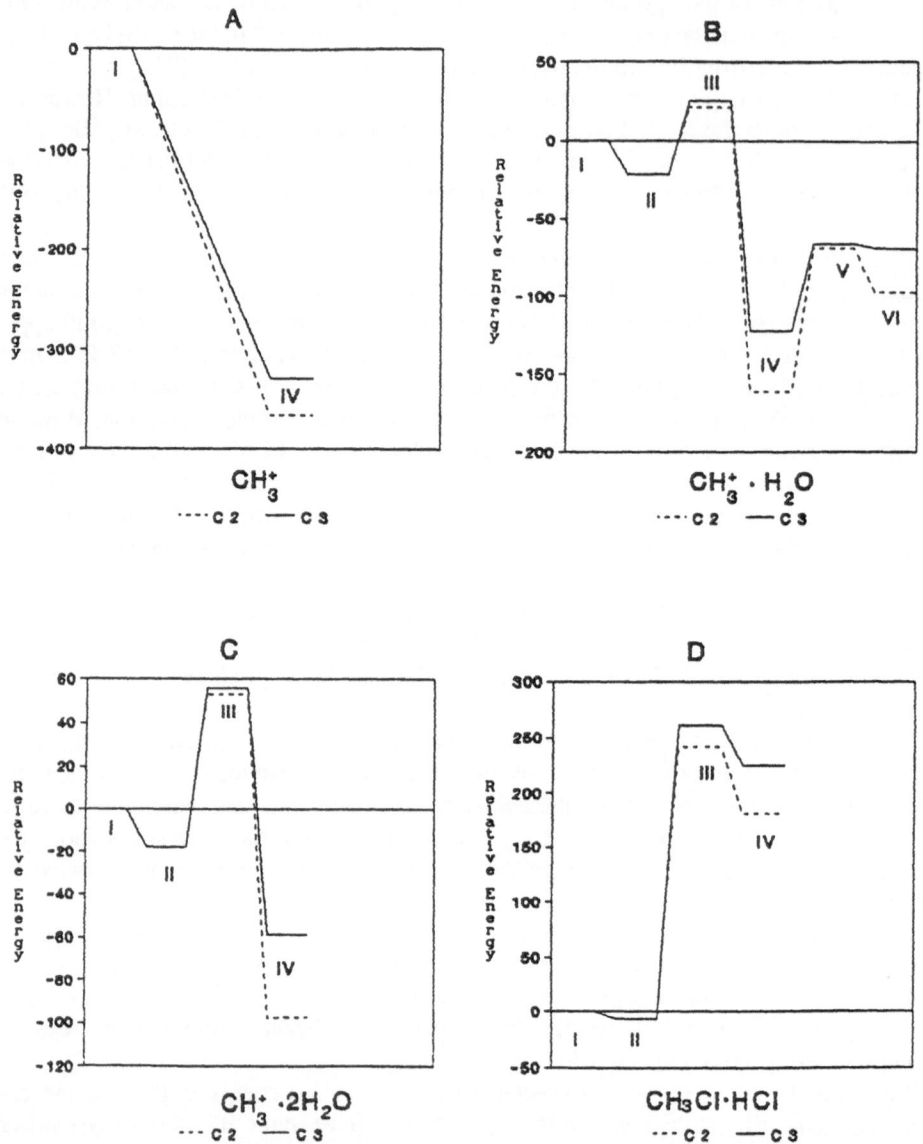

Fig. 3. Calculated reaction energy profiles of the four modelled attacks to Furan by electrophiles a) CH_3^+; b) $CH_3^+ - H_2O$; c) $CH_3^+ - 2H_2O$; d) $CH_3Cl - HCl$. I – Isolated Reactants; II – Prereactive Complex; III – Transition State; IV – σ Complex; POC-Postreactive Complex.

This feature is never be observed in typical solution reactions.

Calculations show the behaviour of $CH_3Cl - HCl$ as a relatively weak reactant with a strong site selectivity. The activation energy is large in both ring positions, but 18.3 kJ/mole higher for C_β than C_α. Boltzmann relative probability to surpass such barrier is calculated to be 0.994 (99.4%) for C_α at room temperature. This guarantees an important preference. In this case, transition states are not SN2 like, because they occur later in the reaction coordinate when the leaving group is already too far from the methyl cation, and the configuration of methyl C atom is inverted, like in typical SE transition structures.

A very interesting feature of this process is the charge transfer between the heterocycle in one side and the electrophile with leaving group in the other, corresponding to the transition state. This results from a comparison of total charge *in the furan ring* before the reaction, which is zero, with the integrated charge of the same ring constituent atoms in the transition state. It can be referred to as the *transferred charge* and interpreted as additional evidence of reaction advance in the transition state with respect to σ complex in each site.

In the case of the monohydrated cation $CH_3^+-H_2O$, the transferred charges in both C_α and C_β methylations are 0.201 and 0.211, respectively. A comparison between these values can also be inferred as evidence of the heteroatom influence in the site selectivity. In this case, such a comparison clearly shows a very slight difference between both sites which indicates the predominance of the electrophile over the heterocyclic oxygen as an electron subtracters and explains the low selectivity in the transition state.

A different situation is observed when the electrophile is the weak $CH_3Cl-HCl$, where the calculated transferred charges for C_α and C_β are 0.338 and 0.445, respectively. In this case, two phenomena are observed: i) the amount of transferred charge is significant larger than in the previous cationic case, which denotes a more advanced reaction in the transition state, and ii) the ring substrate structure has a greater influence in the transferred charge when the attack comes from different positions.

One possibility is to consider the transferred charge in the transition state as a *parameter for reactivity*. This confirms that a lower transferred charge is related with less advance in the reaction when the transition state occurs, with longer distance between the electrophile and the ring, a higher reactivity of the electrophile, and a lower selectivity in the site of heterocyclic aromatic ring substitution, due to the fact that its own structural features are not significantly influencing the transition state. Conversely, when the transferred charge in the transition state is large, all of the previous effects are reversed and, consequently, the ring features determine the site of substitution. A similar behaviour could be hypothesized in electrophilic attacks to substituted phenyl rings.

Experimental results in solution show furan ring substitutions limited to the C_α site [17]. On the other hand, the reaction profiles of our cationic methyl (Fig. 3a) and hydrated cationic methyl (Figs. 3b and 3c) theoretical models show a σ complex which is unrealistically more stable than the reactants, as a consequence of our simulation in a quasi gas phase reaction environment. Therefore, these electrophiles are too strong to represent the species in solution, where solvent molecules soften their Lewis acidity, and further hydrations of the cations yield reacting electrophilic species with higher selectivity. This trend is verified by an extrapolation of the series of progressive hydration from Figs. 3a to 3c. To the above reported slight selectivity due to activation energies, corresponding to the mono- and bihydrated transition states (Figs. 3b and 3c), the important indication of C_α substitution preference due to its σ complexes stability respect to C_β must be added. However, this can only be taken as a quasi static approach, because such σ complexes appear as isolable chemical species.

Among our modelled reactions, the one that resembles a solution like behaviour is that with the weak $CH_3Cl-HCl$ electrophile because the σ complex is less stable than the reactants. In this case the transition state energy (Fig. 3d) is more similar to that of the σ complex, and depends largely on the attack position, favoring C_α.

Considering this evidence, it seems easier to do σ complex theoretical models with soft electrophiles in order to study selectivity and other features when the behaviour in solution is desired to be reproduced. This saves time later when modelling the complete reaction path in larger systems.

It is very well known [2, 3] that electrophilic substitutions in aromatic systems are driven by the determining step in σ complex. However, in the case of the monohydrated methyl cation, the further steps of proton elimination have also been modelled (Fig. 3b). The complete reaction path gives a new transition state before the final products with a new activation energy which appears much lower than the previous one. This confirms that the product formation step does not influence the reaction speed.

Chou and Weinstein [18] made a pioneering theoretical study of the electrophilic attack to the furan ring. They concluded that the selectivity to C_α site is due to the electrostatic potential of heterocyclic oxygen which drives the first approach of the electrophile and favors the C_α substitution, due to its proximity.

As shown above, our calculations of strong electrophiles (ions) include some sort of previous electrostatic complex, which we call prereactive complex (see Figs. 3b and 3c, and Fig. 1a) and is similar to the one reported by Chou and Weinstein. However, in the case of strong electrophiles like the mono and bihydrated methyl cation, energy differences between the one driving to C_α and the one to C_β (Figs. 3b and 3c) are negligible. This could be taken as evidence of the true origin of selectivity, which depends on the nature of transition state and σ complex stabilities, as we have stated earlier in this paper.

Moreover, if the electrophile is as weak as $CH_3Cl-HCl$, where polarization occurs during the reaction path, then the prereactive complex occurs by one of the carbon atoms of the ring in adjacent position with respect to oxygen (Fig. 2a), probably due to the high HOMO population [19]. On the other hand, distances between reactants in such prereactive states appear too great and too similar to the ring dimensions to be able to drive site selectivities.

1.2. THE INFLUENCE OF ENVIRONMENTAL MOLECULES

The specific solvent influence is taken into account by initially modelling the CH_3^+ attack alone and thereafter the same with one and two degrees of solvation, sequentially. Water is taken as a widespread solvent considering both simplicity in calculations and the ubiquitous character of this compound.

Table II shows the mono and bihydration AM1 energies in repect to the isolated CH_3^+ by both C_α and C_β sites. Transition states of the isolated cation are non existent.

In the isolated reactant stage, hydration energies are maximum due to the high localization charge in the methyl cation, and consequently its very low LUMO eigenvalue (see values above). However, transition states are less stabilized by water, and the σ complexes appear more independent of solvation. This can be explained by the weak charge localization in any atom and the considerable increment of LUMO eigenvalue in both cases with respect to the isolated cation.

As expected, the reaction path profile changes significantly with hydration (see Figs. 3a–3c); the increase in the activation barrier and the decrease in the relative stability of the

TABLE II

Solvent influence in methylation of Furan. AM1 hydration energies of reaction species in kJ/mole.

Species		$CH_3^+ \cdot H_2O$	$CH_3^+ \cdot 2H_2O$
Isolated reactants		−229.5	−295.0
C_α	Transition state	−40.2	−74.4
	Sigma complex	−23.5	−25.7
C_β	Transition state	−41.1	−75.8
	Sigma complex	−23.9	−25.9

σ complexes is obvious, as mentioned above.

An increment in the value of the activation energy follows from the significantly higher stability of reactants, from the isolated methyl cation to the bihydrated species. Table II shows a stabilization of 229.5 kJ/mole for the first water molecule in reactants. The second water molecule increases the stability of reactants in additional 65.5 kJ/mole. However, the transition state, responsible for the activation energy, only stabilizes with one water molecule in 40.2 and 41.1 kJ/mole by C_α and C_β, respectively, and with an additional water molecule 34.2 and 34.7 kJ/mole, respectively. Do the theoretical similarity of C_α and C_β transition states as stated before, and the consequent small influence of solvent on site selectivity.

Another inference from the previous arguments is that the theoretical model predicts the overall decrease in the reaction rate, with respect to gas phase, with the electrophile cation solvation due to their increasing activation barrier. The trend shows a realistic view of experimental reacting systems like the methylation of aromatic rings in aqueous alcohols with strong mineral acids, where additional water molecules from the environment can increase the relative energy of the σ complex and effect the actual behaviour [20, 21]. It is a typical case where the environmental molecules modulate the electrophilic character of the main reactant.

Another situation arises with the theoretically modelled $CH_3Cl−HCl$ system, where the actual electrophilic agent is formed during the reaction in the transition state, at the expense of the environmental HCl molecule, which constitutes a protic and polarizing 'solvent'. Here the activation barrier is decreased because of the polarization capacity of the environmental HCl. This effect in chlorination (a very similar reaction to methylation) is experimentally confirmed [22].

1.3. CORRELATION ENERGY CORRECTIONS

Table III shows the energy variations in respect to the prereactive complex obtained by different calculation procedures. The first column is referred to *ab initio* SCF at the split valence 4-31G basis set level, the second includes the MP2 correlation energy correction in the total energy account (calculational details and references in the Appendix), the third is the simple semiempirical valence electrons SCF AM1 values, and the fourth gives the separate correlation energy MP2 correction.

TABLE III

Energy variations for C_α species related to the prereactive complex at *ab initio* 4-31G SCF, 4-31G SCP+MP2, and AM1 levles, and the MP2 correlation energy differences (in kJ/mole) in the $C_4H_4O + CH_3^+ - H_2O$ system by C_α site.

Critical point	ΔE_{SCF}	$\Delta E_{SCF+MP2}$	ΔE_{AM1}	ΔE_{MP2}
Transition state	81.5	36.1	43.8	−45.5
σ complex	−127.5	−94.9	−139.2	32.5

Accurate calculations stabilize the transition state, while the sigma complex is destabilized. Such a smoothing effect on the reaction path profile is also observed with respect to the SCF calculations, which have been performed, as can be seen, with a fine double zeta basis functions for valence electrons which are responsible of bonding interactions. This means that the Hartree-Fock average potential field, also calculated using a feasibly good basis function, is insufficient to reproduce certain approaching effects in chemical reactivity models. In these cases, the electron correlation from the Hartree-Fock average field certainly needs to be taken into account; likewise when a large basis set is being used in relatively extensive calculations.

Table III shows an interesting comparison between the 7.7 kJ/mole transition state energies calculated by both a time consuming split valence basis *ab initio* SCF procedure with correlation energy corrections and the more simple semiempirical and fully parametrized AM1 method. The modelled transition state could be considered as a supermolecule where London energy effects are more significant in certain long interatomic distances than bonding effects. As stated above, such cases are much more sensible to correlation energy corrections.

It is clear that AM1 method, when it is optimized for long distance effects like hydrogen bonds [15] in respect to experimental results of standard systems, implies the correlation energy in its artificial core-core repulsion term (which uses classical gaussian dependences of the energy with respect to the distance matrix) and corresponding parameters.

The σ complex is a molecular cationic species where the AM1 method results are very similar to SCF calculations and not to values including correlation energies.

2. Effects of Substitutions *out* and *in* the Ring

If we consider that substituents can severely perturb the electronic configuration of a typical ring, like benzene, the effect in a precarious aromatic system, like an unsaturated five membered heterocyclic ring must be more important, but unpredictable, due to the partial lacking of the high symmetry almost always present in aromatic hydrocarbons. This pertains to substituents attached to one of the carbon atoms in the ring, which exist *outside* of it and only taking part in its electronic structure by an *exo* bond.

Such heteroaromatic systems also exhibit a substituent *in* the ring, which is the heteroatom itself. This has a dramatic influence on the electronic structure and all of the related chemical reactions.

Therefore, it seems reasonable to model the effects of different substituent in het-

III   IV   V

eroaromatic rings, not only by taking into account the substituent outside the ring, but also those inside the ring. What follows is a treatment of such effects with traditional groups as external substituent, and O, S, and NH as internals in five membered unsaturated rings. Instead of using benzene as the aromatic system reference, Cyclopentadienyl anion is used. However, because of the ionic character of this ring, very few experiments have been performed with it. Thus, a comparison among perturbed systems, specifically with reference to the above mentioned heterocycles, is warranted. Only one previous work appeared [1] to perform theoretical models related to this important reactions, and it was restricted to MNDO calculations of the pyrrole (I) and N-methyl pyrrole protonation.

The competition for protonation between α' and β' ring sites of five membered unsaturated α substituted heterocycles like pyrrole (III), furan (IV), and thiophene (V) has been widely discussed in many papers [11, 18, 19, 22–27]. The α' site electrophilic substitution is always preferred. It has been modelled previously (see above) for unsubstituted furan. This is because the electron population in the highest occupied molecular frontier orbital (HOMO) favours this position over all others [19]. Experimental quantitative evidence can be found in the reported results [28] of different partial rates of substitution to α' and β' sites in substituted furan, benzofuran, thiophene, and benzothiophene. However, it is expected that other factors influence the first steps of the reaction process.

The action of α carbon atom substituents other than H on the preferred further substitutions in α' and β' sites has not been studied experimentally very often. However, a rich discussion about the influence of different α substituents in furan, pyrrole and thiophene on additional substitutions by the α' site exists [28]. Derivatives which favor β' substitutions in pyrrole and thiophene are known, but in the case of furan such influence is only reported in complexes of Lewis acids as $AlCl_3$ with furfural (considered as a furan ring with a formyl group in the α site) [29].

We want to outline the theoretical modelling of the α site substituent influence on the reactivity of other positions. The calculated values are the single molecule gas phase protonation enthalpy in each of such sites according to the heats of formation which are given by MNDO method. Proton affinities (PA) in each one of the protonated sites have been calculated by the thermochemical routine from the MNDO heats of formation in both fully geometry optimized reactants and the σ complex products. Hence,

$$PA = \Delta H_f(\text{heterocycle}) + \Delta H_f(\text{proton}) - \Delta H_f(\sigma \text{ complex}) .$$

The heat of formation of a free proton has been taken as the experimental value of 1536.5 kJ/mole [30]. Thus PA is the heat of the reaction given a σ complex, like those shown in

TABLE IV
MNDO theoretical proton affinities of pyrrole derivatives[a].

R_α	C_β	$C_{\beta'}$	$C_{\alpha'}$
$-NH_2$	898.15	840.96	903.32
$-OCH_3$	879.05	840.01	882.43
$-OH$	867.96	833.07	872.58
$-CH_2CH_3$	852.62	844.62	859.79
$-CH_3$	849.67	942.24	856.71
$-H$	839.19	839.19	847.42
$-F$	813.60	809.80	820.90
$-Cl$	812.37	811.91	819.49
$-CHO$	795.54	809.78	802.21
$-COOH$	669.68	805.38	794.49
$-NO_2$	605.47	752.56	733.50

[a] kJ/mole.

Figure 1c for furan, with the opposite sign. We depart from the consideration that gas phase σ complexes must be similar to those in solution as described in Section 1.

SCF-MO MNDO method must be effective for calculations of proton affinities in the compounds under study, because previous reported values for non substituted rings agreed fairly well with experimental results [6]. Some known limitations of MNDO method, like dealing with neighboring non bonding interactions where hydrogen atoms are involved [31–33], and overestimations of steric hinderings between hydrogen atoms during dihedral angle optimizations [34] are clearly non significant to obtain the desired results in this work.

As a measure of the site selectivity, we have taken the difference between the proton affinities (ΔPA) in both α' and β' ring carbon atoms:

$$\Delta PA = PA_{\alpha'} - PA_{\beta'} .$$

Table IV shows PA results for pyrrole and α substituted derivatives. To test the trends of the substituent action in this ring system in respect to that in benzene, a correlation is done in Figures 4 and 5 between calculated PA's and Hammett's σ constants for each of them [35]. Figure 4 correlates σ_p constants vs. PA's in the α' site, and Figure 5 in the σ_m vs. PA's in the β' site. The correlation for pyrrole is good in both cases: σ_p can be predicted from the PA's by the MNDO theoretical method with $r = 0.979$ and an average error of 0.080; σ_m correlates to $r = 0.946$ and an average error of 0.071.

Table V shows a list of the highest occupied molecular orbital (HOMO) eigenvalues of the neutral monosubstituted molecules calculated by the MNDO method. Figure 6 reflects the general trend between the calculated maximum proton affinity of substituted pyrrole (always by α' except for $-COH$, $-COOH$, and $-NO_2$ by β', see Table IV) and the maximum HOMO eigenvalue of respective neutral molecules. In this case, the high correlation coefficient of 0.987 and the non significant average error of only 5.5 kJ/mole, gives a good chance to predict the ability of overall protonation from the calculation of

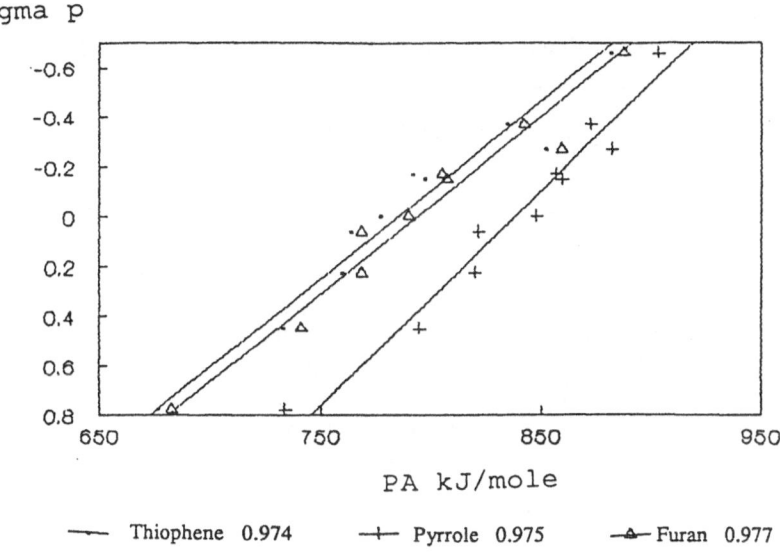

Fig. 4. Correlation of proton affinity by $C_{\alpha'}$ site vs. σ_p.

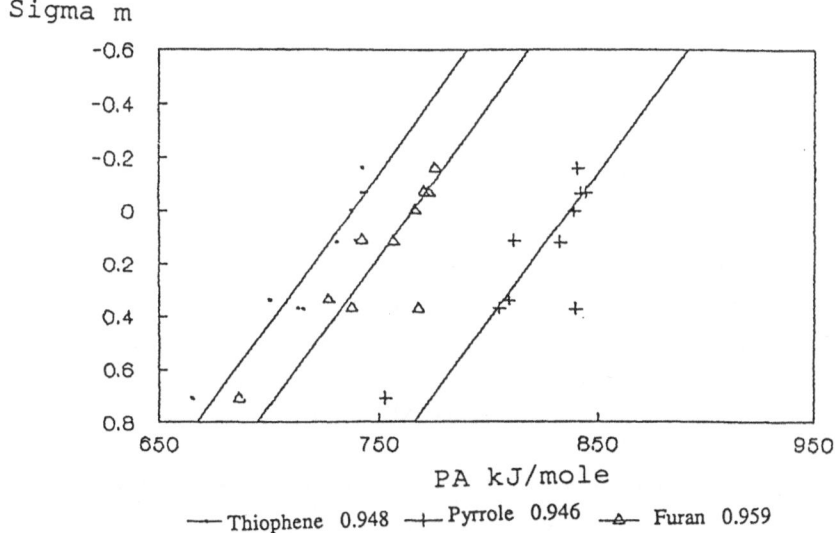

Fig. 5. Correlation of proton affinity by $C_{\beta'}$ site vs. σ_m.

HOMO eigenvalues of the simple neutral substituted molecule. This correlation also reflects the influence of substituents on the frontier orbital.

Position selectivity of pyrrole is illustrated in Table VI by the ΔPA values. From the

TABLE V
MNDO SCF eigenvalues of highest occupied molecular orbitals[a].

R_α	Pyrrole	Furan	Tiophene
$-NH_2$	−7.7152	−9.0970	−8.8637
$-OCH_3$	−8.1925	−8.7210	−9.3400
$-OH$	−8.3996	−8.7377	−9.0015
$-CH_2CH_3$	−8.4940	−9.0220	−9.3510
$-CH_3$	−8.4890	−9.0270	−9.3764
$-H$	−8.5650	−9.1410	−9.5140
$-F$	−8.7960	−9.3706	−9.6184
$-Cl$	−8.8596	−9.3894	−9.6858
$-CHO$	−9.0426	−9.5575	−9.9080
$-COOH$	−9.1296	−9.6739	−9.9729
$-NO_2$	−9.7772	−10.3247	−10.5768

[a] kJ/mole.

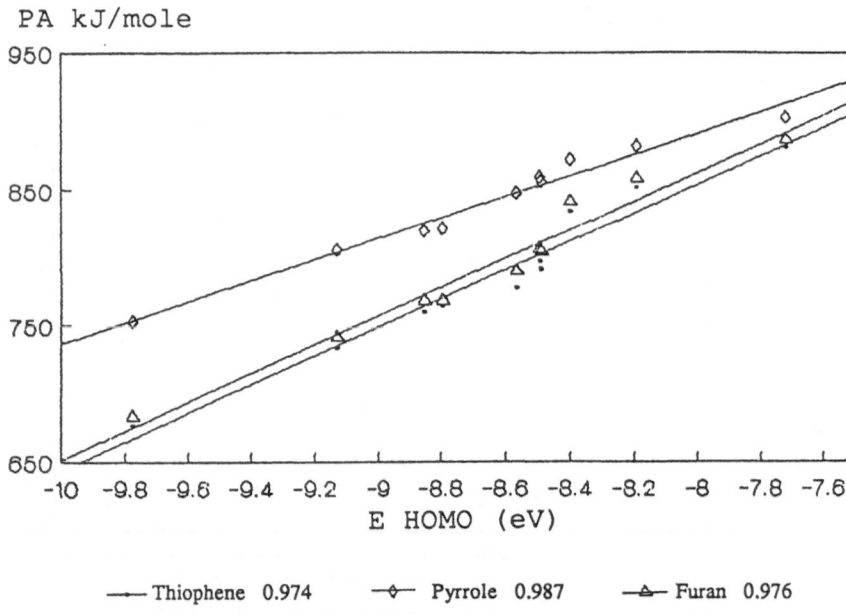

Fig. 6. Correlation between proton affinity and maximum HOMO eigenvalue.

point of view of our theoretical model, all positive differences favor protonation in the α' site. Experimental evidence favors an electrophilic substitution in β' for 2 acetyl pyrrole [36] which is quite similar to 2 pyrrolaldehyde. Therefore, the MNDO prediction can reproduce the site selectivity for all kinds of α substituted pyrroles.

TABLE VI
Difference between MNDO theoretical proton affinity[a].

R_α	Pyrrole	Furan	Tiophene
$-NH_2$	62.36	112.40	140.19
$-OCH_3$	42.42	91.10	112.80
$-OH$	39.51	85.53	103.86
$-CH_2CH_3$	15.17	36.24	52.84
$-CH_3$	14.47	31.55	49.46
$-H$	8.23	22.98	40.12
$-F$	11.10	41.50	63.25
$-Cl$	7.58	26.22	44.26
$-CHO$	-7.57	23.17	40.12
$-COOH$	-10.89	3.65	20.23
$-NO_2$	-19.06	-3.28	11.49

[a] kJ/mole.

A multiple regression fit of ΔPA vs. the Swain and Lupton [37] constants denotes the resonance effects as more important by a factor of 71.6/17.4 in respect to field effects of substituents. This conclusion can be reached because the correlation is significant as $r = 0.980$ and the standard error of estimation is 5.6 kJ/mole. This result also validates the quality of the model in respect to purely experimental substituent constants.

The previous discussion and tables show that substituents affect all unsubstituted sites, but to a greater extent the α' and β positions. This can be verified by the larger dispersion of PA values in these sites in respect to β' one. If we consider the heterocyclic nitrogen, which could be considered as an internal substituent, being evidently a π donating heteroatom, it must promote a higher reactivity to the adjacent α' site. Both, activating substituent and π donor heteroatom effects, at the same time, reinforce the overall preference of the electrophilic attack to this position. The β site is activated by the substituent, and less by the heteroatom. Consequently, it is not favored for a reaction in normal competitive conditions. Martinez reports [19] that SCF-MO atomic net charges, according to both valence electron semiempirical and non empirical methods, give β and β' positions a preferred electrostatic attraction in respect to α or α' sites.

The reverse effect of ring deactivation occurs with electron withdrawing substituents like $-NO_2$ and $-COOH$. The analysis in this case corresponds to the classical one of the meta substitution in benzene: electron clouds over α' and β sites are more deactivated than β', and electrophiles must attack this one preferently. However, it must be taken into account that the heteroatom is always present as an internal substituent, and the deactivating effect of the external one is softened by the *resonance activating nature* of it.

Table VII shows PA's for furan and substituted compounds. The behavior is similar to that in pyrrole in Table IV. There are two main differences:

- The experimental overall reactivity of furan is less than that of pyrrole [38]. According to our theoretical model, PA values are always lower, which agree with experiment. HOMO energies of furan and derivatives (Table V) are also lower than those of pyrrole

TABLE VII

MNDO theoretical proton affinities of furan derivatives[a].

R_α	C_β	$C_{\beta'}$	$C_{\alpha'}$
$-NH_2$	876.53	775.43	887.83
$-OCH_3$	848.07	768.10	859.20
$-OH$	828.53	756.36	841.89
$-CH_2CH_3$	786.11	770.31	806.55
$-CH_3$	783.78	772.73	804.28
$-H$	766.49	766.49	789.47
$-F$	748.23	726.73	768.23
$-Cl$	747.44	742.00	768.22
$-CHO$	730.09	744.24	752.58
$-COOH$	720.33	737.33	740.98
$-NO_2$	657.78	686.38	683.10
$-CHOAlCl_3$	- - - -	653.75	635.74

[a] kJ/mole.

derivatives. This can also be interpreted as an internal substituent effect, in terms of the larger localization of the O heteroatom π contribution, which increases the cis-dienic character of the perturbed a_2 symmetry HOMO and consequently the localized HOMO electron density in α' carbon atom.

– A simple comparison between ΔPA's in Table VI shows that site selectivity for α' of furan is larger than that of pyrrole.

In the case of the $-NO_2$ group, possible β' orientation falls below the numeric expected accuracy of our theoretical model [39]. However, Belinkii reported the experimental preference to β' site halogenations in ALCl$_3$-furfural complexes [40], and this was confirmed by our calculated ΔPA of -18.01 kJ/mole (not appearing in Table VI). Previously, in another published paper [13], we addressed the influence of the electrophilic agent and compared different substituents in attacks by β' site. This paper clarified the differences between protonation and halogenation and helps to interpret the quality of this result.

Figures 4 and 5 show correlations between Hammett's σ_p and σ_m vs. PA's of furans. The slope is clearly different than that of pyrrole, but the correspondence with substituent trends is also fair to $r = 0.977$ and 0.959, and average errors of 0.071 and 0.063 respectively. In this case the multiple regression of ΔPA vs. the Swain and Lupton constants for resonance and field terms favors the resonance effect 2.3 times more (by a ratio of 119.75/12.59, $r = 0.975$ and a standard error of estimation of 9.8 kJ/mole) than the relation obtained for pyrrole.

Figure 6 also illustrates the relationship between the maximum PA and HOMO eigenvalues of furan compounds. The correlation is $r = 0.9883$ and the average error 6.4 kJ/mole. Predictions of protonation ability from the neutral substituted molecule calculations of HOMO eigenvalues are also possible in this case.

Table VIII shows PA's for thiophene and substituted compounds. The experimental

TABLE VIII

MNDO theoretical proton affinities of thiophene derivatives[a].

R_α	C_β	$C_{\beta'}$	$C_{\alpha'}$
$-NH_2$	856.68	741.62	881.81
$-OCH_3$	825.65	739.03	851.38
$-OH$	805.87	730.24	834.10
$-CH_2CH_3$	759.70	743.88	796.72
$-CH_3$	753.41	741.83	791.29
$-H$	736.84	736.84	776.96
$-F$	728.25	700.32	763.57
$-Cl$	721.48	715.04	759.30
$-CHO$	699.45	711.50	739.23
$-COOH$	694.98	712.79	733.02
$-NO_2$	526.11	664.99	676.48

[a] kJ/mole.

behavior is equivalent to that of furan, but a bit less reactive [38]. Therefore, derivatives of PA's are systematically lower than those of furan, which agree with experiment. HOMO energies of thiophene and derivatives (Table V) are also lower than those of both pyrrole and furan derivatives. This can be interpreted now in terms of the highest energy in the series for the S heteroatom π contribution, which increases the cis-dienic character of the common perturbed a_2 symmetry HOMO and consequently the localized HOMO electron density in α' carbon atom.

However, although simple comparisons between ΔPA's in Table VI show that site selectivity for α' of thiophene is larger than that of furan. This contradicts experimental results both in solution [38] and in gas phase [6], which gives a significant increase, but not predominance, in the rate of β' substitution in the case of thiophene. This disagreement can be explained in terms of an overestimation of the substituent effect, in respect to the availability of vacant d levels in the heteroatom. It is known that MNDO is not parameterized for d orbitals, and must include its effects in other parameters. AM1 [15] or PM3 [41] calculations improve on this result because of their announced better parameterization for sulphur.

Figures 4 and 5 show correlations between Hammett's σ_p and σ_m vs. PA's of thiophene derivatives. The slope is clearly different than that of pyrrole and almost the same as furan, and the correspondence with the substituent trend is also fair to $r = 0.974$ and 0.948 respectively, and average errors of 0.071 for both. In this case the multiple regression of ΔPA vs. the Swain and Lupton constants for resonance and field effects favors 2.1 times more (by a ratio of 116.05/13.20, $r = 0.946$ and a standard error of estimation of 14.6 kJ/mole) the resonance effect than the relation obtained for pyrrole, and is less to 0.9 times the one obtained for furan.

Figure 6 also illustrates the relation between the maximum PA and HOMO eigenvalues of thiophene compounds. The correlation is $r = 0.989$ and the average error 6.2 kJ/mole. This is very similar to the values of furan (see above). Therefore, in this case, predictions

TABLE IX

Experimental[a] vs. calculated relative rates of trifluoracetylation of Thiophene and derivatives.

Substituent	k/k_0[b]	
	Exp.	Calc.
$-OCH_3$	1.8×10^6	2.7×10^6
$-CH_3$	3.8×10^2	9.5×10^1
$-C_2H_5$	5.2×10^2	2.4×10^2
$-H$	1.0	8.3
$-Cl$	5.8×10^{-1}	4.1×10^{-1}

[a] $T = 348$ K, in dichloro ethane.
[b] Ratio of rate constants for substitution in α' related to unsubstituted compound.

are also possible of protonation ability from the neutral substituted molecule calculations of HOMO eigenvalues.

Evidently the MNDO theoretical model of thiophene reactivity is not as good as the one for pyrrole and furan. The reason, as it has been pointed out above, must be the known defects in MNDO's parameterization of Sulphur. However, the availability of experimental results related to the rates of trifluoroacetylation [28], in the case of this ring system, allows a correlation with calculated PA's, if we consider that the influence of substituent is equivalent for all electrophilic reactions. In such a case, the theoretical protonation step can be related to activation in this reaction. The Arrhenius exponential model gives a correlation coefficient of $r = 0.975$ and an average error of estimation of 0.436. If we take into account that relative rates range from 1.8×10^6 to 5.8×10^{-1}, then the average error value is in excellent agreement with , and a fair confirmation for the quality of, our theoretical model. See Table IX.

This paper was authored by both theoretical chemists (JRAI and LAM) and experimental chemists (RM) who are specialized in the field of furan derivative polymerization chemistry. We feel that make-up of our research team (this ratio between theoreticians and experimentalists) was a great asset in problem solving and brainstorming. The cationic initiated polymerization of vinyl furan and other related compounds was the result of one of these conversations.

3. Case Study: Polymerization of Alkenyl Furanes

In our previously published article cited earlier [19], we essentially searched for nucleophilic reactivity patterns among different sites of 2-vinylfuran (VIc), 2-isopropenylfuran (VIb) and other vinyl substituted furan monomers. The study was made using a static approach for which any molecular orbital pattern was deemed useful. A good predictor of such reactivity was found using the highest occupied molecular orbital (HOMO) electron densities, as mentioned above.

An explanation for the abnormal behavior of these compounds [42, 43] in respect to the benzenic analogous: styrene (VIIa) and α-methylstyrene (VIIb) was sought. The cationic polymerizations of VIIa and VIc differ dramatically. The propagation step of VIIa always occurs from the vinyl substituent, but VIc and VId actually produce a pseudo-

R$_1$

α' $\overset{O}{\underset{\beta'}{\underset{1}{\overset{}{\underset{\beta}{\bigcirc}}}}}$ —CH = CH$_2$ (c)

R$_1$= —H   (a)
R$_1$= —CH$_3$ (b)
R$_1$= —CH≡CH   (d)
          |
          CH$_3$

VI

R$_2$

m $\overset{}{\underset{p}{\bigcirc}}$ —CH = CH$_2$ (a)

R$_2$= —CH≡CH$_2$ (b)
          |
          CH$_3$

VII

copolymerization, in which $C_{\alpha'}$ in the furan ring takes part and the vinyl group behaves competitively, as usual. On the other hand, at 253 K, polystyrene [poly VIIa] shows no change in structure, whereas the poly VIc acts principally as poly [1-(2-furyl)ethylene], confirming that the propagation takes place mainly through the $C_{\alpha'}$ of the furan ring. However, VId produces the classical vinyl sequence at 195 K showing that, in this case, ring alkylation is completely negligible.

Finding an explanation for the experimental evidence which suggests the role of transfer agent for added 2-methyl furan (VIb) during the cationic polymerization of VIc and VId is also interesting.

It is well known that the initial step in the cationic polymerization consists of an electrophilic attack at the substrate monomer which acts as a nucleophile, with the corresponding formation of the most stable carbocation. In the case of the above mentioned alkenylfurans, the first approach of the electrophile can occur on both the furan ring and the vinyl double bond [42].

In the first part of this paper, the calculated results of the above mentioned substrate monomers are shown and those of other test molecules: VIc, VId, VIb, VIIa and furan (VIa) itself. Part two consists of the calculation of the more plausible structures of the intermediate species models in the gas phase by the same procedure. They have been conceived as the protonation products of the object substrates in the previously (part one) calculated sites of higher reactivity.

As in previous examples, the MNDO [34] method was selected. Environmental effects, which could be important in the course of the reactions are not considered. However, in this case available experimental data were obtained in conditions very close to inert media. This fact supports the comparisons made because the solvation component can be considered as equivalent for all members of the studied reactions.

The intermediate species model geometries are similar to those of σ complexes in Figs. 1c and 2c and were obtained by full optimization of proposed protonated monomers, assuming that they are in the minimum energy of the gas phase reaction coordinate [29]. The intermediate species model geometries have been optimized from the substrates assuming a tetrahedral configuration in the carbon atom which receives the proton. Proton affinities were calculated as in Section 2 above.

TABLE X
HOMO electron densities in reactive sites.

	Furan ring site		Vinyl site		I.P.
	α'	β'	2	1	(ev)
VIc	.25	.08	.18	.07	8.68
VId	.24	.07	.19	.08	8.65
VIb	.35	.15	–	–	9.02
VIa	.36	.14	–	–	9.14
VIIa	–	–	.21	.10	8.83

Results in all of the cases studied gave an optimized $H^+ - C_{\alpha'}$ bond length of 1.11 Å and $H^+ - C_{\alpha'} - C_{\beta'}$ bond angle of 113° for intermediate species models in which proton enters into furan $C_{\alpha'}$. The corresponding values for protonation of C_2 in vinyl group are 1.11 Å for $H^+ - C_2$ bond length and 111° for $H^+ - C_2 - C_1$ bond angle. The $C - H$ bond length could be slightly overestimated because MNDO gives usually the same value for many well-known shorter tetrahedral $C - H$ bonds in hydrocarbons.

Previous semiempirical [44] and *ab initio* [45] calculations attempted to explane the large difference in reactivity between furan ring C_α (or $C_{\alpha'}$) and C_β (or $C_{\beta'}$) positions. Certain other papers tried to explain the problem with many different arguments [46, 47].

In our explanation in Sections 1 and 2 above and in the already cited [19] paper we referred the HOMO electron density calculated values, which agreed better with the reactivities of the experimental target compounds.

These results are shown in Table X and closely resemble those found by using more quantitative methods such as MNDO.

It is well known that ionization potentials (IP), i.e. eigenvalues of frontier orbitals according to the Koopman's theorem, indicate some order of reactivity [48]. This means that the accessibility of electrophilic reactants to lower IP HOMO's is easier than those of higher values. In our case, the MNDO method is known to be parametrized to obtain Koopman's ionization potentials. Table X also shows these values which give the same order of reactivities as indicated in references [43, 49], i.e.: VId > VIc > VIIa. However, values for VIc and VId can be considered as equivalent.

Table XI shows the MNDO results obtained for the proton affinities and electron densities of the lowest unoccupied molecular frontier orbital (LUMO) of intermediate species. LUMO electron densities can be taken, in turn, as an index of further electrophilic character for the first step species in the polymerization reaction.

The calculated proton affinities of the neutral species in vinyl group C_2 confirm the previous order of reactivity predicted by the Koopman's IP: VId > VIc > VIIa. On the other hand, differences between proton affinities of both possible activated complexes of the same neutral species is not significant for VIc (8.2 kJ/mole), slightly larger for VId (16.9 kJ/mole) and clearly larger for VIIa (73.2 kJ/mole). Some deviation with respect to polymerization behavior of VIc at low temperatures is observed, which shows a clear preference for $C_{\alpha'}$ reactivity. However, 8.2 kJ/mole is below the expected accuracy of

TABLE XI

MNDO calculated proton affinities and LUMO electron densities of intermediate species.

Species		P.A.[a]	LUMO electron densities (site)			
$H^+ - C_{\alpha'}$	VIa	786.6	$(C_{\beta'})$	.49	(C_β)	.25
$H^+ - C_{\alpha'}$	VIb	803.3	$(C_{\beta'})$	.28	(C_β)	.01
$H^+ - C_{\alpha'}$	VIc[b]	828.4	(C_2)	.21	$(C_{\beta'})$	.21
$H^+ - C_2$	VIc[b]	836.8	(C_1)	.34	$(C_{\alpha'})$	.29
$H^+ - C_{\alpha'}$	VId	836.6	(C_2)	.21	$(C_{\beta'})$	.20
$H^+ - C_2$	VId	853.5	(C_1)	.37	$(C_{\alpha'})$	.24
$H^+ - C_m$	VIIa[c]	759.4	- - -			
$II^+ - C_2$	VIIa	832.6	(C_1)	.42	(C_p)	.17

[a] In kJ/mole.
[b] Is referred to activated complexes of VIc protonated in α' position of furan ring and 2 of the vinyl bond respectively.
[c] Is referred to activated complex of VIIa protonated in the meta position of phenyl ring.

MNDO for building a theoretical model

The results of ionization potentials (Table X) and proton affinities could explain why VIb can not be protonated in the first step (VIIIa) of the polymerization when VIc and VId coexist in the reaction medium, since their protonation is more favored. This also explains why the oligomerization of VIb in the presence of these monomers is not observed and why more drastic experimental conditons are necessary to obtain it, as has been verified.

On the other hand, it is known that VIb competes advantageously with VIc and VId in the propagation step (IX), and 90% of the alkylation product is obtained in the transfer reaction X. This fact indicates a larger nucleophilicity at this stage with respect to other monomers. At first sight, this seems contradictory with what was said before, but we are looking for its clue in our current research.

As it was said above, MNDO proton affinities are, however, unable to explain why the protonation occurs at the ring $C_{\alpha'}$ site in the low temperature polymerization of VIc. Calculations give a larger value for the attack at the C_2 site (see Table XI). On the other hand, the results are in good agreement with the experimental fact that the low temperature reaction of VId behaves as a typical case of cationic vinyl polymerization.

More accurate proton affinities can be obtained from 4-31G *ab initio* SCF calculations for both alkenylfurans, at a single point with the MNDO optimized geometry. They are shown in Table XII. Results are in very good agreement with previous experimental evidences. The more active positions appear to be $C_{\alpha'}$ in VIc and C_2 in VId, but in the last one differences between C_2 and $C_{\alpha'}$ are less important than those obtained from MNDO results. We can conclude that the semiempirical method underestimated the stability of the furan type carbocation.

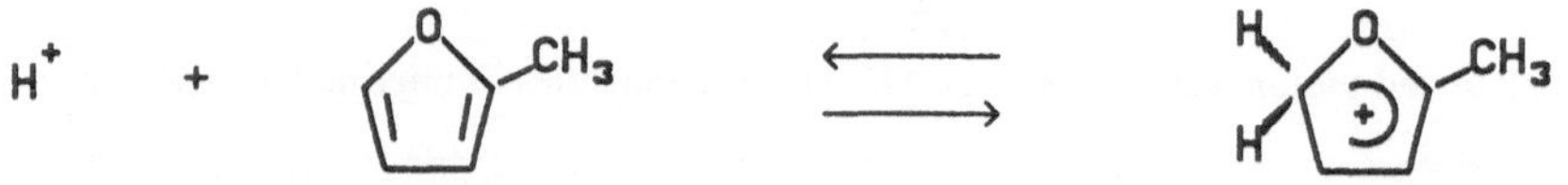

H^+ + [furan with CH$_3$] ⇄ [protonated furan with CH$_3$]   VIIIa

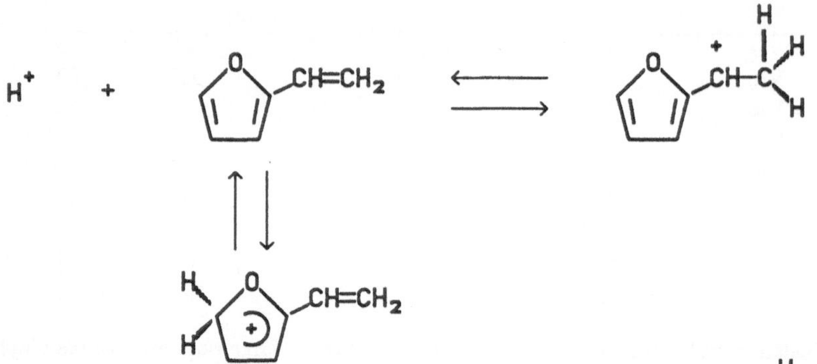

H^+ + [furan with CH=CH$_2$] ⇄ [furan with CH-C cation]   VIIIb

⇅

[protonated furan with CH=CH$_2$]

M^+ + [furan with CH=CH$_2$] ⇄ [furan with CH-C-M cation]

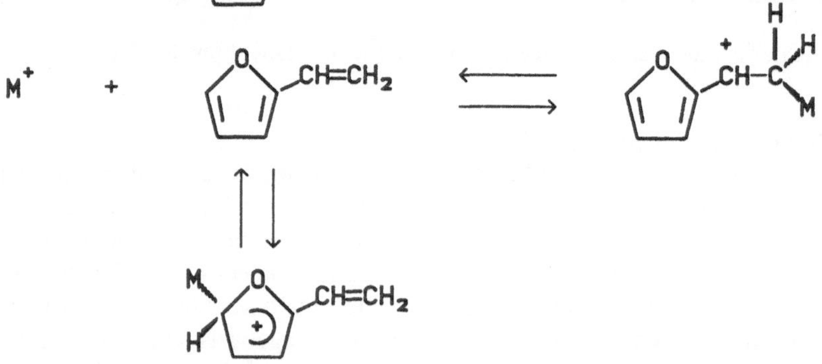

IX

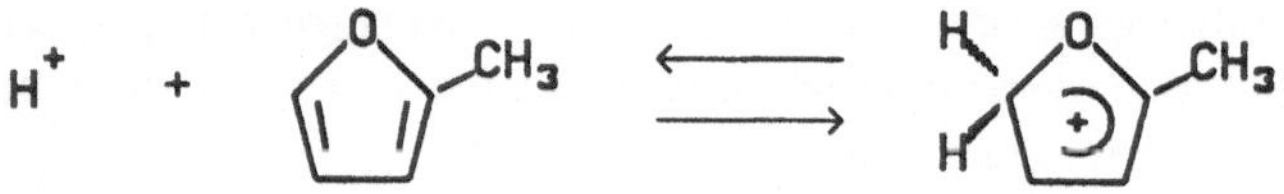

⇅

[furan with M and CH=CH$_2$]

H^+ + [furan with CH$_3$] ⇄ [protonated furan with CH$_3$]   X

4. Conclusions

As stated in the initial paragraph, the purpose of this paper is to illustrate a methodology. Chemists are looking for a theoretical model to help them understand processes and, consequently, to predict behaviours and structures. We have attempted to apply a rather simple methodology to a difficult and complex problem in heterocycle chemistry: the reactivity of furan. The computer programs which were used to perform the calculations made in this paper (perhaps with the exception of MP2 correlation energy correction) are currently available for use on personal computers. They have been used in our laboratory in Havana and can be in most laboratories, even at home (for more information, see Appendix).

The procedure has been found a more suitable method for creating a theoretical model. The science has been to obtain a clearer idea about what is really happening experimentally. We have summarized our findings as follows:

1. AM1 method calculations accurately reproduce many details of the electrophilic substitution reaction path in furan ring methylation, which has been proposed from exper-

TABLE XII

Ab initio 4-31G proton affinities and LUMO electron densities of alkenyl furans.

Species		P.A.[a]	LUMO electron densities (site)			
$H^+ - C_{\alpha'}$	VIc	933[b]	(C_2)	.19	($C_{\beta'}$)	.18
$H^+ - C_2$	VIc	915[b]	(C_1)	.27	($C_{\alpha'}$)	.21
$H^+ - C_{\alpha'}$	VId	951	(C_2)	.20	($C_{\beta'}$)	.19
$H^+ - C_2$	VId	955	(C_1)	.30	($C_{\alpha'}$)	.20

[a] Energies in kJ/mole.
[b] Is referred to activated complexes of VIc protonated in the position α' of furan ring and β of the vinyl bond respectively.

imental results and is also able to model the reaction mechanisms.

2. The results obtained confirm previous supposals about the site selectivity, which appears dependent on the qualities of the transition state, and its energy proximity with the σ complex. Calculated activation barriers for different sites show a general trend to increase C_α site substitution at room and lower temperatures; mixed site substitutions are obtained at higher temperatures.

3. The furan methylation mechanism corresponds to a typical electrophilic substitution with selectivity effects.

4. The *transferred charge* between the substrate ring and the electrophile appears to be a fair indication of the reaction advance in the calculated saddle point, and useful for predicting ring selectivity and electrophile strength.

5. The polar solvent influence in cations appears as not affecting site selectivity because the own structural features of the substrates in both C_α and C_β substitutions are not significant to it.

6. If the electrophilic agent is a previously formed cation, polar solvents appear to raise the activation energy due to the solvation, thus having a stabilizing effect on the strength of the cation.

7. Correlation energy drops the activation energy and rises the σ complex energy. Such an effect trends to smooth the reaction path profile.

8. Reactivity for electrophilic attacks in α substituted rings is demonstrated to be consistent with the HOMO features, which gives a simple first approach to such phenomena in these compounds. The important point of the good correlation between HOMO eigenvalues and proton affinities is its confirmation of previous theoretical considerations [19] about the predominance of the orbital control on the electrophilic substitution of five membered heterocyclic unsaturated rings.

9. From the above discussed results, it can be concluded that in the series of heterocycles III-V, pyrrole results as the most reactive and, additionally, presents the lowest site selectivity, given by the minor ΔPA between α' and β' sites.

10. According to the theoretical results, the electrophilic substitution is favored because the σ complex by the α' site is more stable. Electrodonor substituents direct further substitutions to α' site, while electroacceptors diminish reactivity in both α' and β sites. Therefore, the α'/β' selectivity is diminished by increasing the electroacceptor

character of the α substituent. The more important donation and subtraction of electrons occurs by resonance (π symmetry) effects.

11. The σ complex stability diminishes in the series pyrrole > furan > thiophene, while the energy difference between α' and β' complexes changes in the reverse manner.

12. From the simple MNDO gas phase theoretical model, it is possible to make a prediction of substitutions, and to estimate the influence of a substituent on the relative rate of electrophilic substitutions.

13. Correlations with Swain and Lupton constants show that the resonance effect is predominant in all ring systems. However, pyrrole exhibits a relatively greater incidence of field effects or a lower incidence of resonance effects. This clearly indicates the different nature of heterocyclic N in pyrrole, in respect to O in furan and S in thiophene, for enhancing the transmission of electronic cloud perturbations to other molecular sites in another manner.

14. Initiation mechanism of cationic polymerization of vinyl monomers can be modeled as an attack at the more reactive sites in the target molecule. In our case, they appear to be C_2 in the vinyl substituent, for both phenyl and furan compounds. However, the pentagonal oxygenated ring also contributes the competitive nucleophilic $C_{\alpha'}$ site for this process.

15. Both *ab initio* 4-31G and MNDO calculations on 2 iso propenyl furan (VId) confirm that the typical vinylic propagation of the cationic polymerization is more favoured in it than in 2 vinyl furan (VIc), from both the energy of the intermediate species and HOMO electron density points of view.

Acknowledgements

We would like to thank the many people who have contributed to this work. Thanks are due to a small team of undergraduate students from the Faculty of Chemistry of the Universidad de La Habana, which included Diana Cuza, María Caridad González, Rita María Isoba, who performed many of the calculations and prepared many of the input sets. Professors Carlos Pérez and Pedro Ortiz in Havana, were an enormous influence on the work as a whole. Professor Yves G. Smeyers, Dr. Alfonso Hernández (both in Madrid), and Dr. R.R. Musin (Novosibirsk) were very helpful in explaining their latest research. We would also like to thank our host institutions: the Institute of Chemical Kinetics and Combustion of the Siberian Branch of the Academy of Science of the USSR in Novosibirsk, the Instituto de Estructura de la Materia (CSIC) in Madrid, and the Department of Quantum Chemistry of the Uppsala University in Uppsala, all of whom provided outstanding technical support. Finally, thanks are due to the University of Havana (Havana), the Academy of Sciences of the USSR (Moscow), the Consejo Superior de Investigaciones Científicas (Madrid), UNESCO (Paris), and SAREC (Stockholm) for generous financial support.

Appendix: Methods

As mentioned in the main part of the paper, semiempirical SCF-MO procedures AM1 [15] and MNDO [34] have been used to optimize all molecular geometries and to obtain approximate values of the reaction coordinate energetics. The programs used are a current version of AMPAC adapted for microcomputers [16]; MNDO, a program of the TC-HABANA package of theoretical chemistry in microcomputers [50].

Minimization methods to obtain optimized geometries are those built in such programs, like Davidon-Fletcher-Powell's, where first derivatives of energy respect to each internal coordinate to be optimized (gradients) are found after each SCF loop and changes in all of such independent variables are done during each optimization cycle towards mutual relaxation of such gradients up to a predefined limit. Transition states have been found to be the same using the built-in Powell's gradient norm minimization method, looking for metastable atomic configurations. That is a state in which all of the second derivatives of energy in respect to internal coordinates are positive, except one. This is the so-called *saddle point* in the hypersurface. Powell's method looks for a near to equivalent situation where gradient norm is zero, or near to zero.

In some of the studied reactions and structures, previously obtained AM1 or MNDO optimized geometries have been used to carry out *ab initio* SCF MO calculations with the standard 4-31G basis set [51] and MP2 procedure contained in the GAUSSIAN 80 program [52].

Since this paper was written, more TC-HABANA [50] programs have become available to perform many of the above calculations using personal computers. For example, HAVPAC [53] can be used to calculate MNDO, AM1, and PM3 optimized geometries, and ENHANCED MICROMOL [53] can be used to perform *ab initio* calculations. Please contact the authors for further information.

References

[1] Nalewajski, R.F., Koninski, M.: *J. Molec. Struct. (THEOCHEM)* 1988, **165**, 365.
[2] Melander, L.: *Acta Chem. Scan.* 1949, **3**, 95.
[3] Bonner, T.G., Bowyer, F. and Williams, G.: *J. Chem. Soc.* 1953, 2650.
[4] Olah, G.A.: *J. Am. Chem. Soc.* 1965, **87**, 1103.
[5] Olah, G.A. and Kiovsky, T.E.: *J. Am. Chem. Soc.* 1967, **89**, 5692.
[6] Houriet, R., Schwarz, H. and Zummack, W.: *Nouveau J. Chem.* 1981, **5**, 10.
[7] Angelini, G., Lilla, G. and Speranza, M.: *J. Am. Chem. Soc.* 1982, **104**, 7092.
[8] Laguzzi, G. and Speranza, M.: *J. Chem. Soc. Perkin Trans. II* 1987, 857.
[9] González, J.T. and Sordo, J.T.: *J. Mol. Struct.* 1985, **120**, 3.
[10] Carey, F.A. and Sundberg, R.J.: Advanced Organic Chemistry. Part A: Structure and Mechanisms, Plenum Press, New York, 1977.
[11] Alvarez, J.R., Montero, L.A. and Martinez, R.: *Makromol. Chem.* 1990, **191**, 281.
[12] Alvarez-Idaboy, J.R., Cuza, D., Montero, L.A. and Isoba, R.: *J. Molec. Struct. (THEOCHEM)* 1990, **209**, 361.
[13] Alvarez-Idaboy, J.R., González, M.C. and Montero, L.A.: *Folia Chim. Theoret. Latina* 1989, **17**, 39.
[14] Hammond, G.S.: *J. Am. Chem. Soc.* 1955, **77**, 334.
[15] Dewar, M.J.S., Zoebisch, E.G., Healy, E.F. and Stewart, J.J.P.: *J. Am. Chem. Soc.* 1985, **107**, 3902.
[16] Liotard, D.: AMPAC program, PC version 3.0, Austin, Texas, 1988.
[17] Katritsky, A.R. and Lagowski, J.M.: The principles of Heterocyclic Chemistry, Academic Press, New York, 1968.
[18] Chou, D. and Weinstein, H.: *Tetrahedron* 1978, **34**, 275.
[19] Martinez, R. and Montero, L.A.: *Acta Polymerica* 1987, **38**, 153.
[20] Olah, G.A., Cupas, C.A., Comisarow, M.B. and Pittman Jr., C.U.: *J. Am. Chem. Soc.* 1965, **87**, 2997.
[21] Olah, G.A. and Halpern, Y.: *J. Org. Chem.* 1971, **36**, 2354.
[22] Stock, L.M., Baker, F.W.: *J. Am. Chem. Soc.* 1962, **84**, 1661.
[23] Gorb, L.G. and Abronin, I.A.: *Izv. Akad. Nauk SSSR, Ser. Khim.* 1982, **2**, 342.
[24] Gorb, L.G., Morozova, I.M., Belenkii, L.I. and Abronin, I.A.: *Izv. Akad. Nauk SSSR, Ser. Khim* 1983, **4**, 828.
[25] Gorb, L.G., Abronin, I.A. and Korsunov, V.A.: *Izv. Akad. Nauk SSSR, Ser. Khim* 1984, **5**, 342.

[26] Abronin, I.A.: *Zh. Org. Khim.* 1981, **17**, 1134.
[27] Belenkii, L.I. and Abronin, I.A.: *Zh. Org. Khim.* 1981, **17**, 1134.
[28] Clementi, S., Linda, P. and Marino, G.: *J. Chem. Soc. B* 1971, 79.
[29] Fernandez, J. and Sordo, T.L.: *J. Mol. Struct.* 1985, **120**, 3.
[30] Bowers, M.T.: Gas Phase Ion Chemistry, Academic Press, New York, 1979.
[31] Dannenberg, J.J. and Vinson, L.K.: *J. Phys. Chem.* 1988, **92**, 5635.
[32] Salk, S.H.S., Chen, T.S., Hagen, D.E. and Lutrus, C.K.: *Theor. Chim. Acta* 1986, **70**, 3.
[33] Hamou-Tahra, Z.D., Kassab, E., Allavena, M. and Evleth, E.M.: *Chem. Phys. Lett.* 1988, **150**, 86.
[34] Dewar, M.J.S. and Thiel, W.: *J. Am. Chem. Soc.* 1977, **99**, 4907.
[35] Hammett, L.P.: Physical Organic Chemistry, McGraw-Hill, New York, 1940.
[36] Suarez, M., Otazo, E. and Pina, M.C.: Introducción a la Química de los Heterociclos, Ed. Universidad de La Habana, Habana, 1988.
[37] Swain, C.G. and Lupton, E.C.: *J. Am. Chem. Soc.* 1968, **90**, 4328.
[38] Marino, G.: *Chim. Ind.* 1973, **55**, 349.
[39] Dewar, M.J.S.: *J. Phys. Chem.* 1985, **89**, 2145.
[40] Belinkii, L.I.: personal communication.
[41] Stewart, J.J.P.: *J. Comp. Chem.* 1989, **10**, 209.
[42] Martinez, R.: *Rev. Cienc. Quim.* 1984, **15**, 1.
[43] Gandini, A. and Martínez, R.: *J. Polym. Sci., Polym. Symp.* 1976, **56**, 79.
[44] Hermann, R.B.: *Int. J. Quantum Chem.* 1968, **2**, 165.
[45] Nguyen, M.T., Hegarty, A.F., Ha, T.K. and De Maré, G.: *J. Chem. Soc. Perkin Trans. 2* 1986, 147.
[46] Houriet, R., Roll, E., Bouchux, G. and Hopillard, Y.: *Helv. Chim. Acta* 1985, **68**, 2037.
[47] Politzer, P. and Weinstein, H.: *Tetrahedron* 1975, **31**, 915.
[48] Epiotis, N.D.: Theory of Organic Reactions, Springer Verlag, Berlin, 1978.
[49] Gandini, A. and Martínez, R.: *Makromol. Chem.* 1983, **184**, 1189.
[50] Montero, L.A., Alvarez, J.R., del Bosque, J.R., González, M.C., Cano, R. and Rodríguez, M.C.: *Folia Chim. Theoret. Latina* 1988, **16**, 33.
[51] Ditchfield, R., Hehre, W.J. and Pople, J.A.: *J. Chem. Phys.* 1971, **54**, 724.
[52] Gaussian 80, QCPE version IBM-MVS, February, 1982.
[53] Montero, L.A. and Alvarex-Idaboy, J.R.: to be published.

SINDO1 STUDY OF THE PHOTOREACTION OF TETRAMETHYLENE SULFIDE

KARL JUG and FRANK NEUMANN
Theoretische Chemie
Univei sität Hannover
Am Kleinen Felde 30
3000 Hannover 1
Germany

ABSTRACT. The mechanism of the photochemical decomposition of tetramethylene sulfide (TMS) was investigated by the semiempirical MO method SINDO1. The relevant singlet and low lying triplet potential energy
hypersurfaces were studied and intermediates and transition structures were optimized with limited configuration
interaction (CI). The predominant yield of ethylene cannot be explained by the decomposition via triplet intermediates which are responsible for most other products. The main ethylene production should be generated from the
fragmentation of the third singlet which undergoes a topological crossing and can be identified as the first excited
singlet in the product region. This pathway could be called adiabatic.

1. Introduction

The study of photochemical reaction mechanisms of heterocyclic compounds has been a
favorite topic in organic photochemistry. In the photoisomerization of heterocyclic aromatic
rings a permutation of ring atoms takes place, usually in a complicated process. Different
mechanisms were proposed for the photochemical rearrangement of different heterocycles
containing nitrogen, oxygen or sulfur [1]. We have studied such rearrangements for unsubstituted and substituted furan [2], pyrrole [3] and thiophene [4] by quantum chemical
methods. Besides rearrangement ring compounds can also undergo decomposition. This is
a particularly important mode of product formation in the photochemistry of saturated rings
without π electrons [5]. The question that arises naturally is which bond is the preferential
site for bond breaking because this is an essential step for product selection. In this context
compounds with sulfur atoms are of special interest because sulfur can be divalent, tetravalent and hexavalent. Its divalent character would indicate similarity of bonding with the
respective oxygen compound, whereas tetravalent sulfur may have similarity with carbon
which is also indicated by the close electronegativity of the two elements. In the case of
hypervalence the influence of d orbitals is a matter of debate. Therefore it seemed to us
necessary and promising to study the following series of five membered rings containing
sulfur: tetramethylene sulfide (TMS), also called tetrahydrothiophene or thiolane, tetramethylene sulfoxide (TMSO) and tetramethylene sulfone (TMSO$_2$). After the review of the
early work [5], several groups [6–8] had undertaken comparative or single experimental

L. A. Montero and Y. G. Smeyers (eds.), Trends in Applied Theoretical Chemistry, 99–113, 1992.

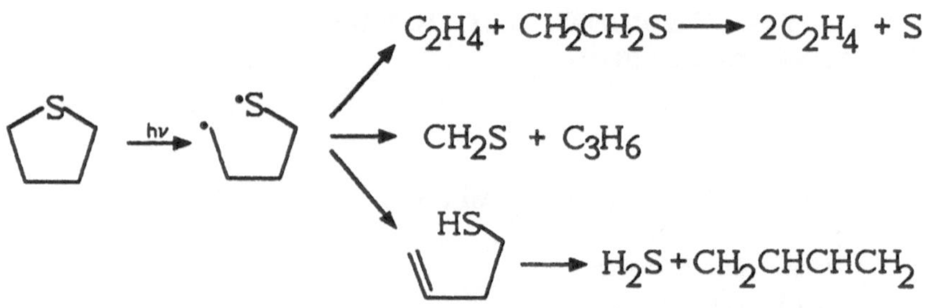

Fig. 1. Reaction mechanism proposed from experiment.

studies on the photochemical mechanisms of these compounds. Their results were challenging in the following respect. Mechanisms for bond breaking and subsequent reactions were established on the basis of the products obtained. However, the product yield cannot always be explained on the basis of these mechanisms. This is particularly true for the photoreaction of TMS. We have therefore undertaken an explicit theoretical study of ground and excited state potential surfaces of this compound. In the following sections we wish to present the results and the conclusion concerning the adopted mechanism.

2. Method of Calculation

The calculations were performed with the semiempirical MO method SINDO1 [9]. The extension to second-row elements includes d orbitals [10] and was tested extensively for sulfur compounds [11] with respect to geometries and binding energies. The suitability of the method for the study of photochemical reaction mechanisms was demonstrated for cyclopentanone [12] besides furan, pyrrole and thiophene [2–4]. In the following, we shall denote the ground state by R_0, the vertically excited singlet states by R_1, R_2 etc. and the triplet states by $^3R_1$ etc.. Intermediates on the third singlet surface will be labelled I_3, on the lowest triplet surface by $^3I_{1a}$, $^3I_{1b}$, $^3I_{1c}$ etc. and transition structures on the ground state surface by TS_0. Finally products are denoted by P_{0a}, P_{0b} etc.. Minima including intermediates are characterized by all positive roots of the force constant matrix, whereas transition structures have one negative root. Ground state geometries for equilibria and transition structures were located by complete geometry optimization with a Newton-Raphson procedure [13, 14]. Bond lengths were optimized within 1% and bond angles and dihedral angles within 0.1°. Excited state structures and transition structures were optimized on their respective CI surfaces with an accuracy slightly less than for the ground state equilibrium structures. Further details of the optimization procedure can be found in the furan treatment [2]. The size of the configuration interaction was adjusted to guarantee a unambiguous qualitative explanation of the mechanism.

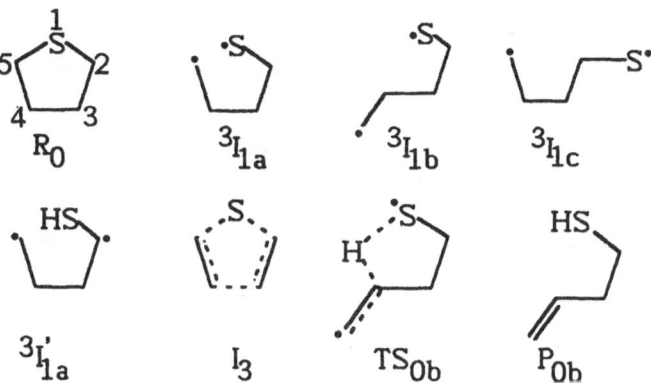

Fig. 2. Geometrical structures of reactant R, intermediates I, transition structure TS and rearranged product P.

3. Results and Discussion

3.1. GEOMETRIES AND BINDING

From the experimental data on tetramethylene sulfide Scala *et al.* [6] deduced the reaction mechanism presented in Figure 1. After excitation the CS ring bond is broken and a diradical formed which is considered as the precusor of three different products: ethylene, cyclopropane or propene and butadiene. No sulfur products were traced. It was therefore essential to locate the important stationary points on the potential hypersurfaces of the system. From previous work on cyclopentanone [12] we expected that the postulated diradical is a minimum on the lowest triplet surface. Besides the cis- or gauche minimum, two different trans minima could exist on the this surface, which are distinguished by the orientation of the radical centers with respect to the rest of the ring. We label this potential minima as $^3I_{1a}$, $^3I_{1b}$ and $^3I_{1c}$ (Figure 2). $^3I_{1a}$ can be the precursor of ethylene and cyclopropane, $^3I_{1b}$ of ethylene, cyclopropane, propene and butadiene, $^3I_{1c}$ of ethylene. Migration of hydrogen from adjacent carbons to sulfur can convert $^3I_{1a}$ to $^3I'_{1a}$ which can be a precursor of ethylene. The experimentalists' scheme in Figure 1 has to be modified to describe the migration of hydrogen for the formation of 4-butenethiol. This can take place only from the trans intermediate $^3I_{1b}$. The transition structure is TS_{0b} and the product P_{0b}. From this product butadiene can be formed by elimination of hydrogen sulfide. A major addition to the experimentalists' scheme is the consideration of an intermediate I_3 on the third singlet surface. This seems to be necessary to explain the high yield of ethylene production. With this frame as the background information the structures of Figure 2 were optimized on their potential surfaces. For the four triplet diradicals $^3I_1$ and the intermediate I_3, configuration interaction (CI) of size 6×6 to 16×16 with single and double excitations was used, for TS_{0b} a 14×14 CI was used and the reactant R_0 and product P_{0b} were optimized on the SCF surface. The CI was selected in such a way that the lowest triplet state and the transition structure were correlated by the appropriate excitations. A criterion for the inclusion of a determinant in the CI calculation for each state was that the coefficient of this determinant contributed more than 0.1% to the considered state.

TABLE I

Bond lengths r (Å) bond angles α and dihedral angles ϕ (degrees) of reactant R, intermediates I, transition structure TS and rearranged product P.

	r_{34}	r_{45}	r_{23}	r_{15}	r_{12}	r_{27}	r_{17}	r_{410}	r_{110}
$R_0(C_{2v})$	1.570	1.566	1.566	1.821	1.821	1.095	2.402	–	–
$I_3(C_{2v})$	1.766	1.450	1.450	1.960	1.960	1.090	–	–	–
$^3I_{1a}$	1.558	1.509	1.552	3.215	1.742	1.110	2.106	–	–
$^3I_{1b}$	1.542	1.523	1.551	4.871	1.733	1.123	2.020	1.089	3.250
$^3I_{1c}$	1.582	1.490	1.528	4.746	1.725	1.148	1.845	–	–
$^3I'_{1a}$	1.543	1.498	1.523	3.676	1.742	2.348	1.347	–	–
TS_{0b}	1.567	1.392	1.560	–	1.781	–	–	1.409	1.880
P_{0b}	1.540	1.345	1.552	4.477	1.844	–	–	2.594	1.343

	α_{345}	α_{234}	α_{123}	α_{327}	α_{217}	α_{2110}	α_{3410}
$R_0(C_{2v})$	112.1	112.1	108.9	111.2	–	–	110.0
$I_3(C_{2v})$	114.2	114.2	106.9	114.7	–	–	–
$^3I_{1a}$	117.8	118.0	120.3	92.5	–	–	–
$^3I_{1b}$	117.7	116.7	122.1	87.3	–	58.5	111.3
$^3I_{1c}$	118.5	121.3	126.3	77.0	–	–	–
$^3I'_{1a}$	120.7	121.6	120.4	–	98.1	–	–
TS_{0b}	125.5	115.3	111.4	110.1	–	87.6	101.9
P_{0b}	131.0	115.4	124.6			101.1	112.0

	ϕ_{5432}	ϕ_{4321}	ϕ_{4327}	ϕ_{3217}	ϕ_{21410}
$R_0(C_{2v})$	0.0	0.0	-119.3	–	–
$I_3(C_{2v})$	0.0	0.0	–	–	–
$^3I_{1a}$	-9.9	-52.5	-160.0	–	–
$^3I_{1b}$	163.4	76.1	182.2	–	150.5
$^3I_{1c}$	3.5	-154.9	-64.4	–	–

	ϕ_{5432}	ϕ_{4321}	ϕ_{4327}	ϕ_{3217}	ϕ_{21410}
$^3I'_{1a}$	-12.3	-68.3	–	172.7	–
TS_{0b}	117.7	-9.3	-130.5	–	180.9
P_{0b}	0.0	0.0	–	–	180.0

The results of this geometry optimization are presented in Table I. Here atom 7 is the hydrogen atom bound to $C^{(2)}$ in R_0 and atom 10 is the hydrogen atom bound to $C^{(4)}$.

To understand the bonding, atomic valencies and bond orders are also presented in Table II. Atom 6 is the other hydrogen atom bound to $C^{(2)}$ in R_0. Atomic valencies were calculated in the CI form [15] of the concept presented by Gopinathan and Jug [16]. Bond orders were calculated by the maximum bond order principle [17] with the projection technique [18]. From the criterion for diradicals [19, 20] it is apparent that the underlined numbers indicate radical centers and classify the five triplet intermediates as diradicals. Their atomic valencies are reduced by approximately one unit compared with centers which have the normal valency close to 4 for carbon and close to 2 for sulfur. For all triplet

TABLE II

Atomic valencies V and bond order P of reactant R, intermediates I, transition structure TS and rearranged product P.

	V_1	V_2	V_3	V_4	V_5	V_7	V_{10}			
$R_0(C_{2v})$	2.227	3.980	3.992	3.992	3.980	0.998	–			
$I_3(C_{2v})$	1.278	3.709	3.660	3.660	3.709	–	–			
$^3I_{1a}$	1.244	3.924	3.929	3.979	3.080	0.934	–			
$^3I_{1b}$	1.265	3.989	3.944	3.965	3.073	0.920	0.928			
$^3I_{1c}$	1.272	3.819	3.904	3.977	3.104	0.905	–			
$^3I'_{1a}$	2.224	3.148	3.932	3.902	3.085	0.996	–			
TS_{0b}	1.795	3.974	3.967	3.883	3.690	–	0.872			
P_{0b}	2.105	3.979	3.994	3.977	3.979	–	1.000			

	P_{34}	P_{45}	P_{23}	P_{15}	P_{12}	P_{26}	P_{27}	P_{17}	P_{410}	P_{110}
$R_0(C_{2v})$	1.203	1.195	1.195	1.211	1.211	0.980	0.980	–	–	–
$I_3(C_{2v})$	0.800	1.526	1.526	0.834	0.834	0.976	0.976	–	–	–
$^3I_{1a}$	1.176	1.390	1.188	0.016	1.391	0.973	0.928	0.163	–	–
$^3I_{1b}$	1.188	1.400	1.196	0.027	1.401	0.973	0.910	0.209	0.953	0.035
$^3I_{1c}$	1.147	1.423	1.198	0.022	1.375	0.973	0.873	0.297	–	–
$^3I'_{1a}$	1.162	1.406	1.299	-0.015	1.389	0.958	–	0.973	–	–
TS_{0b}	1.190	1.853	1.185	–	1.264	–	–	–	0.577	0.677
P_{0b}	1.230	2.116	1.201	–	1.176	–	–	–	0.043	0.978

intermediates the bond $C^{(5)}S$ is broken as can be seen from the bond orders P_{15} which are close to zero. The bond orders P_{34} and P_{23} are important for the fragmentation process. Breaking of $C^{(3)}C^{(4)}$ in the triplet intermediates would result in ethylene, whereas breaking of $C^{(2)}C^{(3)}$ would lead to cyclopropane or propene. From the comparison of P_{34} and P_{23} we see that the $C^{(3)}C^{(4)}$ bond is always weaker. We would therefore expect that the formation of ethylene is more favorable than of cyclopropane or propene. Finally the bond orders P_{410} and P_{110} of TS_{0b} indicate that the migration of hydrogen from carbon to sulfur has progressed slightly more than half way. Different from the other cases I_3 is not a diradical because only the sulfur atomic valency is substantially reduced from its normal value two. The bond orders of I_3 show already the pattern for the subsequent fragmentation in two ethylene and sulfur.

3.2. VERTICAL EXCITATION AND REACTION PATHWAYS

Scala and coworkers [6] irradiated TMS with UV light of 147 nm which is equivalent to 8.44 eV and investigated the fragments by gas chromatography and a mass spectrometry. Quantum yield calculations resulted in 1.06 mol ethylene, 0.073 mol butadiene, 0.064 mol propene, 0.044 mol acetylene and 0.019 mol cyclopropene per mol TMS. Deuterated TMS resulted in the same products. Cyclobutane was not found. Sulfur containing compounds were not included in the analysis.

We calculated the vertical excitations with a 182×182 CI including 156 single excitations from the 6 highest lying occupied orbitals (HOMOs) to the 13 lowest lying unoccupied orbitals (LUMOs), 24 interpair double excitations and one double excitation from HOMO to LUMO. The selection for the single excitations was done in such a way that all close-lying HOMOs beyond the second were included in the calculation until the first large gap of 2 eV from HOMO6 to HOMO7 appeared in the sequence. From the LUMOs all MOs not dominated by d orbital contributions were included. Since TMS belongs to point group C_{2v} we could classify the first singlet excitations by irreducible representation, excitation energy and oscillator strength as R_1 (A_2, 4.72 eV, 0), R_2 (B_2, 5.71 eV, 0.02), R_3 (B_1, 7.77 eV, 0.07), R_4 (A_1, 8.32 eV, 0.001). From comparison with experiment it seems clear that R_3 is the most likely candidate to be efficiently reached by excitation in the UV light range used because the other states like R_4 to R_7 have a much smaller oscillator strength.

In the following we have set up a reaction scheme illustrating the various possibilities for photochemical decomposition of TMS (Figure 3). In the figure we have included also a pathway for simultaneous breaking of the sulfur carbon ring bonds. The results for the energies of the stationary points are obtained with the same 182×182 CI used for the vertical excitations. In this scheme we have labelled and encircled pathways 1, 2, 3, 4a, 4b, 5a, 5b and 6 in the sequence of their kinetic feasibility. The barriers on the relevant potential surfaces increase in this sequence. From the alternative pathways 4a and 5a are more favorable than 4b and 5b, respectively.

We now present the detailed reaction pathways for the formation of products along these pathways in the following figures. Figure 4 shows the production of ethylene and sulfur. The reaction starts with a vertical excitation to R_3 and proceeds via I_3 over a small barrier towards the products. The reaction coordinates are $r_{12} = r_{15}$ and r_{34}. All three bonds break simultaneously. All other coordinates are optimized in C_{2v} symmetry on the lowest B_1 singlet surface along the reaction coordinate. The pathway could be called adiabatic. We have optimized the geometries on those surfaces which are marked by a circle on the potential curve at the relevant point. Figure 5 shows the potential curves for the formation of butadiene. The reaction starts from a vertical excitation to R_3 which is followed by intersystem crossing to the triplet manifold where internal conversion leads to the lowest triplet surface. From symmetry considerations we have concluded that intersystem crossing and internal conversion are both allowed. The first coordinate, the dihedral angle ϕ_{2345} for twisting is relevant for the change between R_0 and $^3I_{1b}$. We used a linear interpolation between these two structures to get the necessary information on the energies in between these two minima. Between $^3I_{1b}$ and P_{0b} the second coordinate, the bond length r_{410} for hydrogen migration becomes important. The transition structure was fully optimized. From P_{0b} elimination of H_2S leads to butadiene.

Figure 6 shows the formation of ethylene via $^3I_{1c}$. The twist angle ϕ_{1234} is relevant between R_0 and $^3I_{1c}$. A linear interpolation is used as in Figure 5 between these two geometries. Beyond $^3I_{1c}$ points were optimized along the reaction coordinate r_{13} until the maximum on the singlet surface was overcome. The transition structure TS is not fully optimized. From TS we proceeded to P_{0c} by optimization along bond length r_{34}.

Figure 7 illustrates the formation of acetylene. From R_0 to $^3I_{1a}$ CS bond r_{15} is broken. Hydrogen migration is described by SH bond length r_{17} and ethenethiol formation by breaking of CC bond r_{34}. Acetylene is then formed by elimination of H_2S from ethenethiole.

Figure 8 shows the formation of cyclopropane. The twisting along dihedral angle ϕ_{2345} is the first step to $^3I_{1b}$, the formation of CC bond r_{35} the second step until the barrier TS

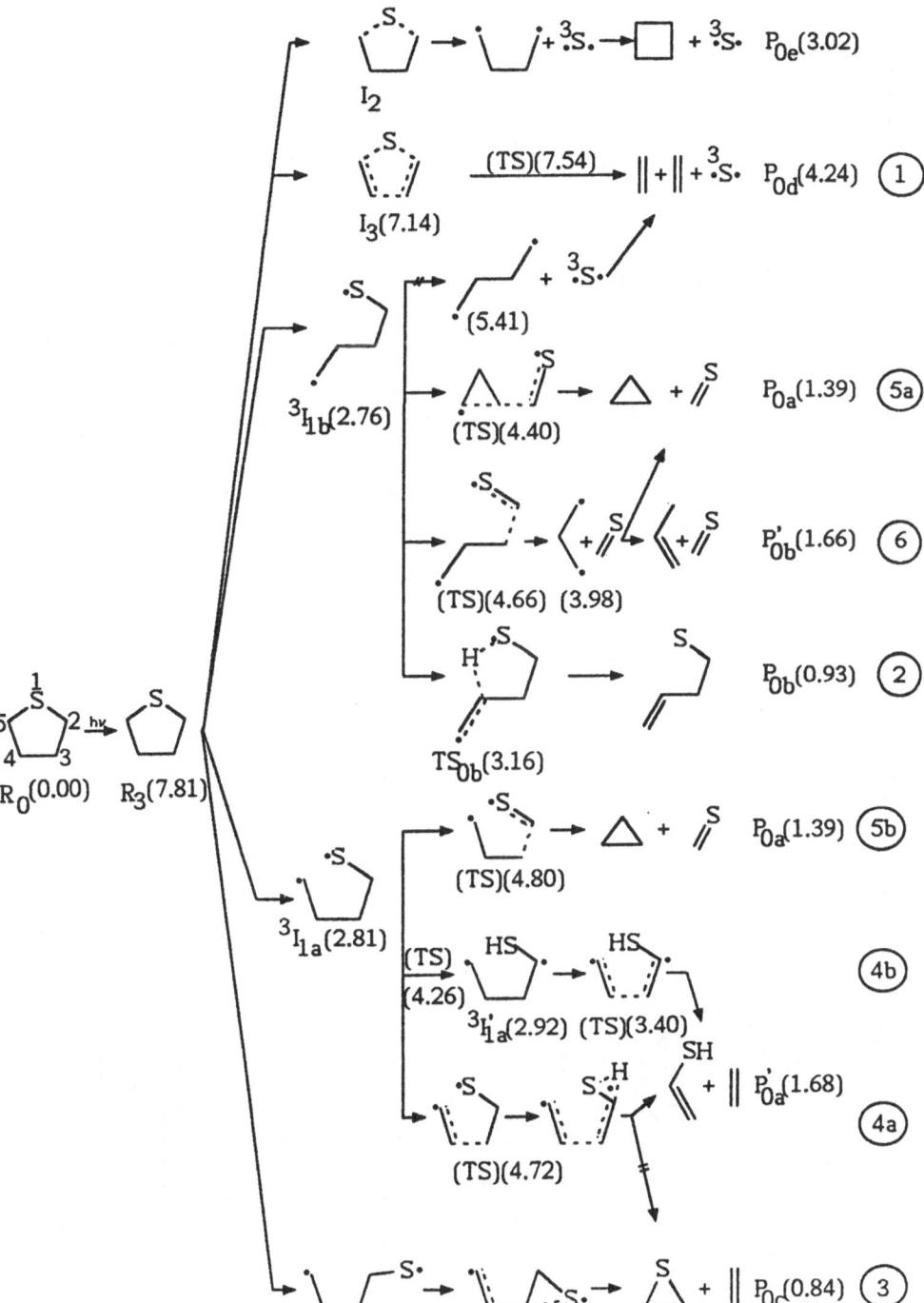

Fig. 3. Reaction scheme.

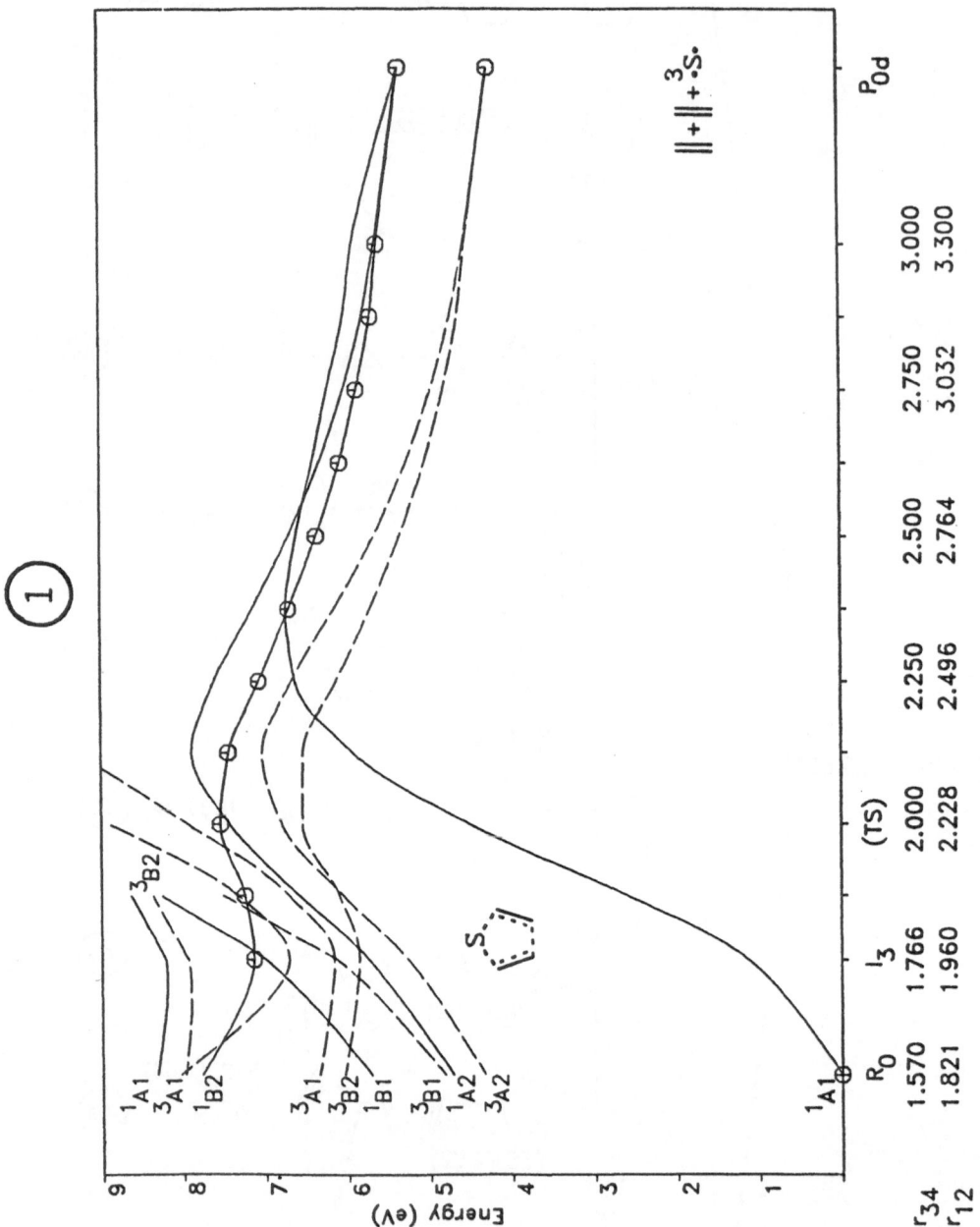

Fig. 4. Potential curves for formation of ethylene via I_3; ◯ points of optimized geometry.

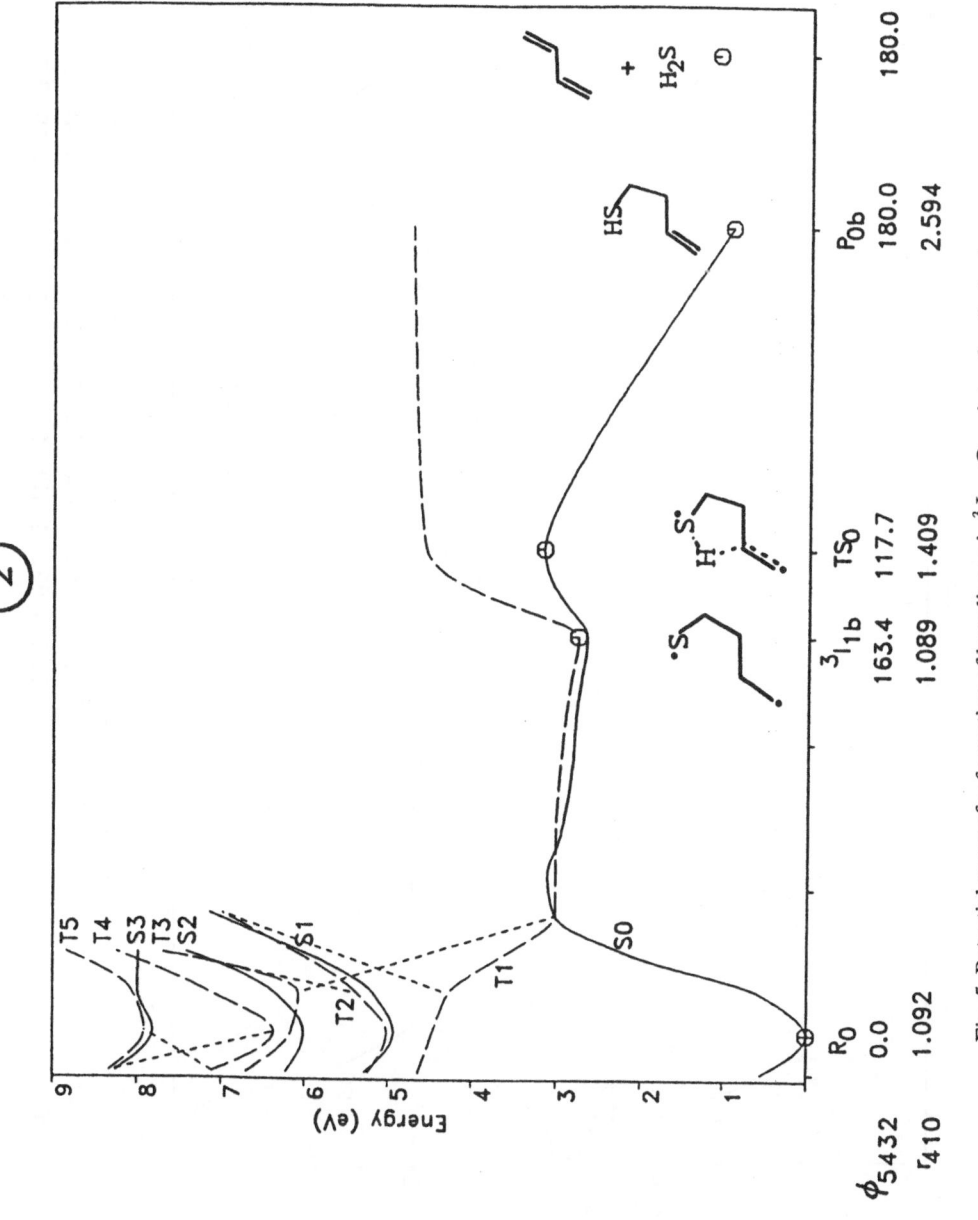

Fig. 5. Potential curves for formation of butadiene via $^3I_{1b}$; $\bigcirc$ points of optimized geometry.

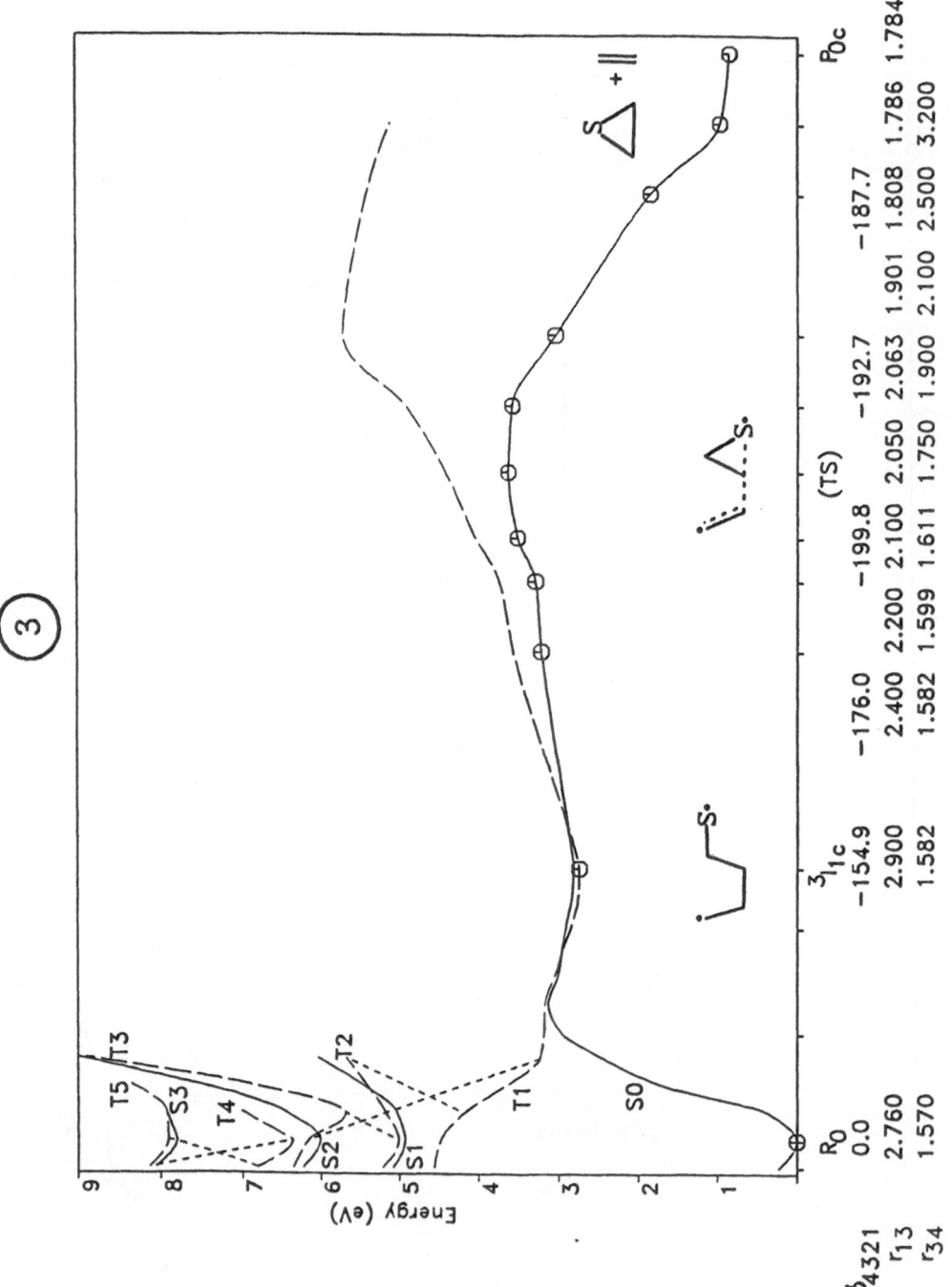

Fig. 6. Potential curves for formation of ethylene via $^3I_{1c}$; ◯ points of optimized geometry.

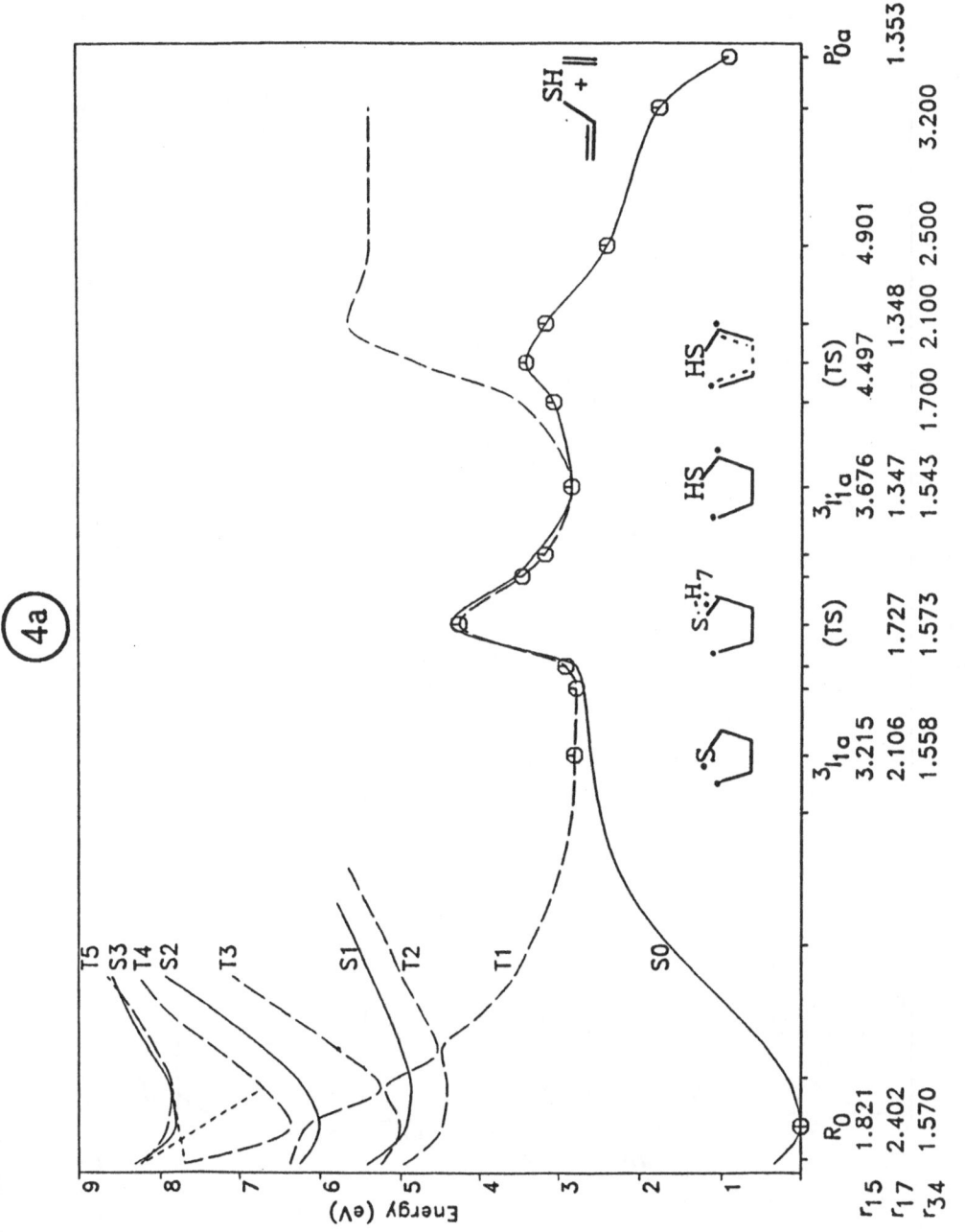

Fig. 7. Potential curves for formation of acetylene via $^3I_{1a}$; ○ points of optimized geometry.

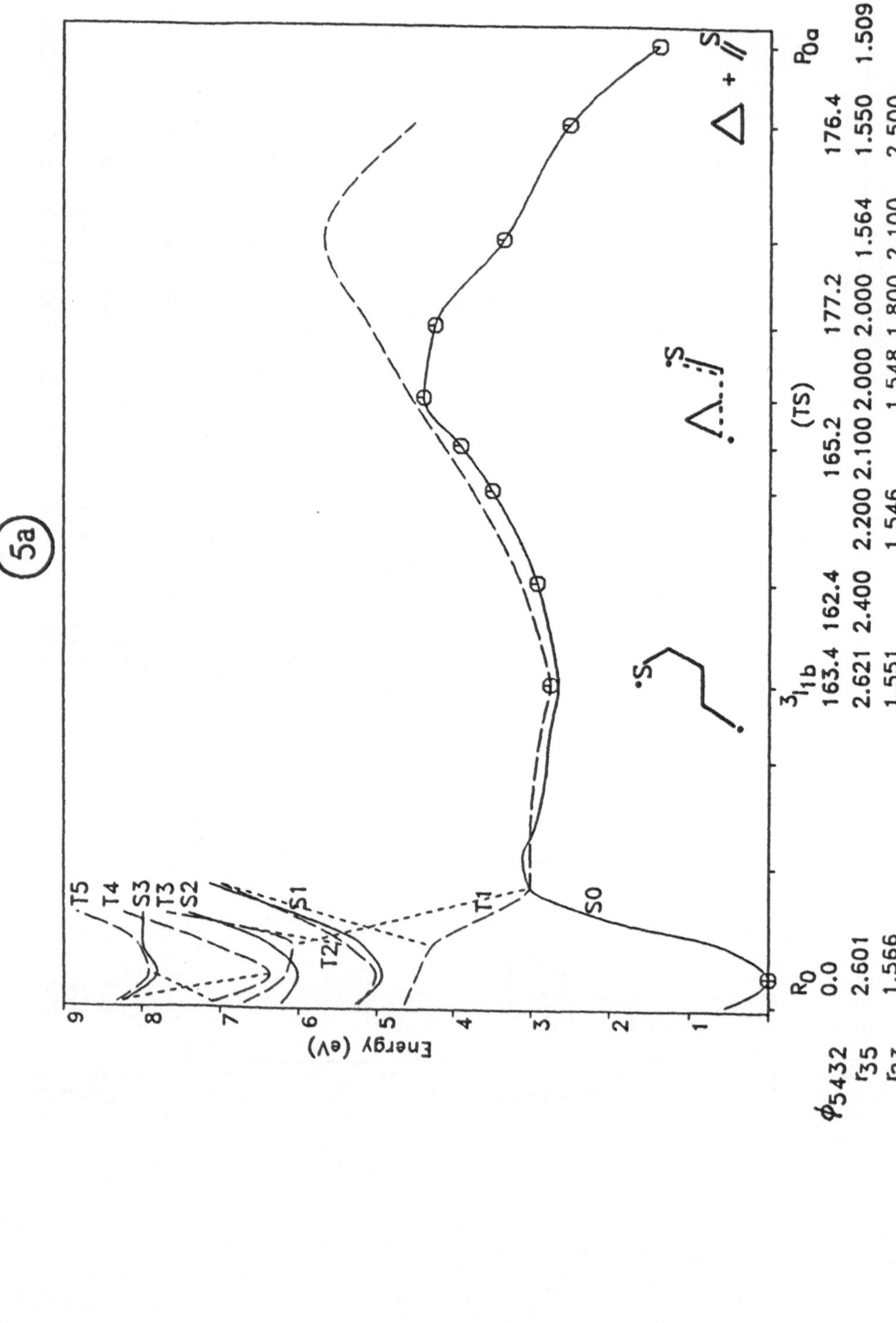

⑤a

Energy (eV)

ϕ_{5432}
r_{35}
r_{23}

	R_0		$^3I_{1b}$	162.4	165.2	(TS)	177.2		176.4	P_{0a}
ϕ_{5432}	0.0		163.4							
r_{35}	2.601		2.621	2.400	2.200	2.100	2.000	2.000	1.550	1.509
r_{23}	1.566		1.551		1.546		1.548	1.564	1.550	2.500
							1.800	2.100		

Fig. 8. Potential curves for formation of cyclopropane via $^3I_{1a}$. ○ points of optimized geometry

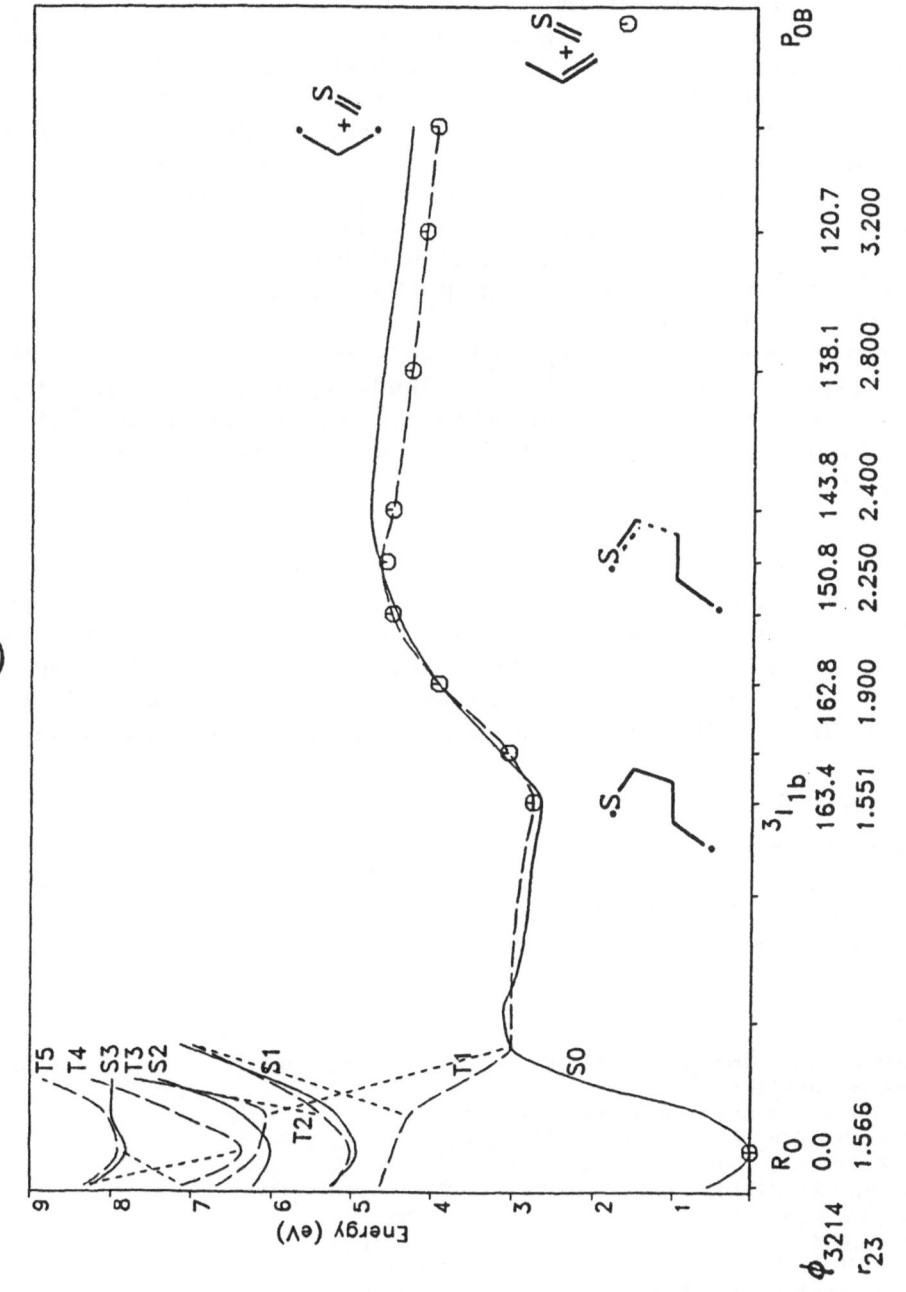

Fig. 9. Potential curves for formation of propene via $^3I_{1b}$; ○ points of optimized geometry.

and finally the breaking of CC bond r_{23}.

Figure 9 shows the formation of trimethylene, which can convert into cyclopropane via ring closing or into propene via hydrogen shift. Between R_0 and $^3 I_{1b}$ the situation is the same as in Figure 8. Trimethylene production is described by breaking CC bond r_{23}.

The relative stability of cyclopropane compared to propene is wrong on the SCF surface. This influences also the sequence of barriers. The sequence of stabilities of cyclopropane and propene can be corrected via CI with double excitations as has been shown before. In this sense, all calculated barriers predict a sequence of product yields in agreement with experiment. Although the barriers for reactions in Figures 4 and 5 are almost equal, the reaction via allowed crossing on the singlet should be faster than via intersystem crossing, thus favoring the former. This would also mean that from the excited TMS molecules few will reach the triplet intermediates $^3 I_1$.

The corresponding reaction starting from R_2 was not investigated here but is included in the reaction scheme. After dissociation of sulfur Braslavsky and Heicklen [21] found cyclobutane, a product which was not present in Scala and Colon's experiments. We conclude that tetramethylene should occur as an intermediate which can either form cyclobutane or dissociate to two ethylenes.

4. Conclusions

An investigation of the photochemical reaction of the saturated ring tetramethylene sulfide brought about the necessity to study seven different reaction pathways related to one intermediate on a singlet surface and three different initial and one subsequent intermediates on the lowest triplet surface. The experimental quantum yields can be explained by single molecule reactions for all products (ethylene, butadiene, acetylene, cyclopropane, propene).

Acknowledgment

Communication with Prof. A. A. Scala on the reaction mechanism is appreciated. The calculations were performed on the CYBER 180/995 and VP 200 EX at Universität Hannover and CRAY X-MP at ZIB Berlin. We thank Deutsche Forschungsgemeinschaft and Fonds der Chemischen Industrie for partial support of this work.

References

[1] Padwa, A.: *Rearrangements in Ground and Excited States*, Vol. 3, p. 501 ff, P. de Mayo, Ed., Academic Press, New York 1980.
[2] Buss, S. and Jug, K.: *J. Am. Chem. Soc.* **109**, 1044 (1987).
[3] Behrens, S. and Jug, K.: *J. Org. Chem.* **55**, 2288 (1990).
[4] Jug, K. and Schluff, H.-P.: *J. Org. Chem.*, in press.
[5] Braslavsky, S. and Heicklen, J.: *Chem. Rev.* **77**, 473 (1977).
[6] Scala, A.A. and Colon, I.: *Dep. Energy ITIC* **15**, 11399 (1978);
    Scala, A.A., Colon, I., and Rourke, I.: *J. Phys. Chem.* **85**, 3603 (1981).
[7] Dorer, F.H. and Salomon, K.E.: *J. Phys. Chem.* **84**, 1902 (1980).
[8] Schuchmann, H.P. and von Sonntag, C.: *J. Photochem.* **22**, 55 (1983).
[9] Nanda, D.N. and Jug, K.: *Theor. Chim. Acta* **57**, 95 (1980).
[10] Jug, K., Iffert, R., and Schulz, J.: *Int. J. Quantum Chem.* **32**, 265 (1987).
[11] Jug, K. and Iffert, R.: *J. Comput. Chem.* **8**, 1004 (1987).
[12] Müller-Remmers, P.L., Mishra, P.C., and Jug, K.: *J. Am. Chem. Soc.* **106**, 2538 (1984).
[13] Jug, K. and Hahn, G.: *J. Comput. Chem.* **4**, 410 (1983).

[14] Himmelblau, D.M.: *Applied Nonlinear Programming*, McGraw-Hill, New York 1972;
     Fletcher, R.: *Practical Methods in Optimization*, Vol. I, J. Wiley & Sons, Chichester 1980.
[15] Jug, K.: *J. Comput. Chem.* **5**, 555 (1984).
[16] Gopinathan, M.S. and Jug, K.: *Theor. Chim. Acta* **63**, 497, 511 (1983).
[17] Jug, K.: *J. Am. Chem. Soc.* **99**, 7800 (1977).
[18] Jug, K.: *Theor. Chim. Acta* **51**, 331 (1979).
[19] Jug, K.: *Tetrahedron Lett.* **26**, 1437 (1985).
[20] Jug, K. and Poredda, A.: *Chem. Phys. Lett.* **171**, 394 (1990).
[21] Braslavsky, S. and Heicklen, J.: *Can. J. Chem.* **49**, 1316 (1971).

THEORETICAL APPROACHES TO STRUCTURAL STABILITY AND SHAPE STABILITY ALONG REACTION PATHS

GUSTAVO A. ARTECA
Department of Chemistry
University of Saskatchewan
Saskatoon
Saskatchewan
Canada S7N 0W0

ABSTRACT. This chapter reviews recent developments on the analysis of interrelations between configurational rearrangements and changes in potential energy and molecular shape. The notions of *structural stability* (associated with energy changes) and *shape stability* (associated with changes in molecular surfaces) are discussed. Procedures used to locate *shape transitions* (that is, the topological equivalent to geometrical structural transitions) along reaction paths or molecular dynamics trajectories are presented and exemplified for some chemical reactions. The techniques provide a useful insight into the changes in molecular properties along reactions paths, something which is not always apparent when studying the potential energy exclusively. The role of the energy changes in modifying the molecular shape is discussed in detail and we provide quantitative measures of structural similarity for the conformations of flexible molecules.

1. Introduction

A theoretical chemist looks at a molecule as an entity provided with *a structure* (represented by the spatial distribution of nuclei), *a shape* (conveyed, for example, by the three-dimensional distribution of the electron density around the nuclei), and *a total energy value* associated with it.

Energetic stability with respect to nuclear rearrangements is the reason for the existence of chemical species. However, energy alone is not a sufficient criterion to characterize molecular properties. The effect of nuclear rearrangements on the electron density is not necessarily the same as on the energy. Moreover, molecular properties such as size and volume are related to the electron density rather than to the energy [1–9]. The interpretation and prediction of these empirical parameters is difficult without introducing the notion of an envelope surface providing molecules with defined shape and size.

This semiclassical view of molecules can be reconciled with quantum mechanics if the geometrical notion of chemical species as fixed configurations is replaced by a topological one. In this latter approach, species are represented by continua of nuclear arrangements (open sets in configuration space). A construction of these continua from the local properties of the potential energy surface has been proposed by Mezey [10]. Bader [11] has proposed another approach where a species can be defined from curvature properties of the electron density function.

L. A. Montero and Y. G. Smeyers (eds.), Trends in Applied Theoretical Chemistry, 115–133, 1992.

The description of chemical species in terms of shape is currently being developed. The same can be said regarding the invariance of shape features when a molecule is allowed to vibrate or rotate. These problems are of concern to several areas of pure and applied science, where one deals with *the dynamics of the molecular interaction*. In pharmacology and molecular biology, for example, the aim is to understand the mechanisms of drug-receptor interactions in order to facilitate a more efficient and rational design of new drugs. In molecular modeling, the knowledge of the active conformations and their persistence over small configurational rearrangements becomes essential. Molecules with comparable biochemical activities are assumed to have similar rigid shapes [12]. But, more importantly, they should possess similar *dynamical shapes* in the neighborhood of the receptor. Synthesis planning and molecular engineering are other fields where interrelations between energy, shape, and conformational rearrangements must be characterized.

In this review we discuss some recent results on the dependence of molecular shape on conformational changes, paying special attention to reaction paths. The techniques used must be adapted to the complexity of the molecular system. For instance, small molecules and macromolecules require different approaches.

2. Geometrical and Topological Concepts of Chemical Structure

Most standard descriptions of molecules in quantum chemistry are solidly entrenched within the validity of the Born-Oppenheimer approximation. In this picture, nuclei are dealt with as classical particles whose positions can be completely specified. This is, of course, in contradiction with the uncertainty principle. As a consequence of this, one should not define chemical species in terms of fixed nuclear configurations [10, 13].

There are a number of alternatives in the literature to incorporate rigorously the quantum mechanical nature of the atomic nuclei into the wavefunctions. A most appealing proposal is the so-called *generator-coordinate approximation* [14]. However, it is debatable whether it is possible to recognize within this context a molecular structure for anything more complicated than diatomic systems. An interesting discussion on the notion of molecular structure in quantum mechanics beyond the Born-Oppenheimer approximation can be found in [15–18].

A different alternative for a satisfactory treatment of the uncertainty in the nuclear positions is provided by molecular topology. In molecular topology one does not move away from the Born-Oppenheimer approximation, but the semiclassical description of the nuclear positions is "corrected" by considering the chemically relevant entities as open sets of nuclear geometries instead of a single configuration.

The topological model gives a description of molecules radically different from the conventional geometrical description. Within the geometrical model, a molecular structure is associated with nuclear geometries. In contrast, nuclear geometries and reaction paths are replaced by a continua of nuclear configurations within the topological approach [10]. Both geometrical and topological approaches use the Born-Oppenheimer approximation. The topological model, however, replaces individual geometries by topological families and, in this sense, it "goes beyond" the semiclassical aspects of the approximation.

As mentioned, the topological classification of nuclear geometries is equivalent to a partitioning of the nuclear configuration space into equivalence classes. The definition of these equivalence classes is flexible. Different properties can be used to this end. For instance, one can collect nuclear geometries in equivalence classes by taking into account the curvature properties of the energy hypersurface (thus defining catchment regions [10]) or by considering the shape of associated molecular surfaces (thus defining the shape invariance regions).

3. Catchment Regions and Shape Invariance Regions

The nuclear configurations associated with an N-atomic molecule can be described with a 3N-dimensional cartesian frame. However, as it is well known, a choice of another space with reduced dimensions is more appropriate to take into account the fact that rigid translations and rotations do not affect the molecular properties (provided that the molecule is isolated and in the absence of external fields).

A convenient choice of space is the so-called configurational space M, with dimensionality $n = 3N - 6$, in general. The elements of this space are not nuclear geometries but equivalence classes, denoted as *configurations K*. A configuration K represents the equivalence class of all nuclear geometries in a $3N$-dimensional cartesian frame which are superimposable (equivalent) to one another by rigid rotations and translations [10]. Reaction paths and conformational rearrangements are represented as parametrized paths in M.

This space M can be partitioned into physical domains by introducing some additional property which defines, in turn, equivalence classes between the configurations K. Depending on the chosen property, these equivalence classes can be given a clear chemical meaning.

A partitioning motivated by the notion of structural stability can be put in terms of the potential energy. Consider a configuration K_0, $K_0 \in M$, representing, for example, a minimum on the potential energy hypersurface. One can define an open neighborhood around point K_0 formed by the collection of all configurations which relax to K_0 by a steepest descent path. This defines the catchment region or basin of the critical point K_0. For a minumum, this neighborhood has dimensionality n. The whole neighborhood represents a collection of configurations which are reversible deformations of K_0. It is reasonable to consider the *whole* catchment region as *the stable minimum*.

When K_0 is a simple saddle point (a critical point of index 1), its catchment region will be an $(n-1)$-dimensional ridge. This catchment region represents the set of configurations which relax reversibly to the saddle point. However, this domain does not represent a stable species since a deformation along the remaining coordinate brings K_0 into the catchment region of another critical point with lower energy.

In summary, one can associate with every critical point K_i (of index λ) a region $C(\lambda, i)$ that is the so-called catchment region of the critical point K_i. The set $\{C(\lambda, i)\}$ is a partitioning the space M:

$$M = \bigcup_{\lambda, i} C(\lambda, i), \tag{1a}$$

$$C(\lambda, i) \cap C(\lambda, j) = \phi, \quad i \neq j. \tag{1b}$$

Using the language of topology, one says that the set $\{C(\lambda,\ i)\}$ defines a subbase for a topology of M. This is the reason why the study of these descriptions of chemical species based on partitionings of configurational space is refered to as *molecular topology* [10].

A partitioning of space M can be formulated by using different principles. For example, one can define regions in configurational space characterized as containing configurations sharing a feature other than the energetic stability. The shape of a molecular surface [13, 19–22] is a property which provides an alternative for classifying nuclear arrangements [23–30].

Molecular shape is a different concept from that of nuclear configurations. A conformational rearrangement, for example, involves continuous changes of the nuclear positions, energy, and the electron density. However, the *essential shape features* of molecular surfaces (as described, for example, by isodensity contours) may remain invariant for some moderate distortions.

The partitioning of space M based on molecular shape [23–29] has some conceptual differences from the partitioning based on an energy criterion. The essential differences are: a) Molecular shape (or for that matter the shape of anything) cannot be uniquely defined. At most, one can decide *which shape features are relevant* and somehow *characterize* them. b) The molecular shape varies depending on which is the property chosen to depict a molecule in 3-space. The description will be different whether one chooses the electron density or van der Waals surfaces (VDWSs) to represent a molecule.

The relation between these two descriptions of the configurational space is exemplified in Figure 1. Here we show a generic potential energy function for a chemical system described over a restricted, working subset M_w of the space M. The energy function has a number of minima and saddle points connecting them. The figure indicates how one can associate an energy value $E(K)$ and a molecular surface $G(K)$ with each configuration K. The changes in energy and in shape features of the surface can be followed over the surface, thus providing a characterization of the space M_w where the configurations are defined.

Consider a molecule in a given nuclear configuration K, described by a molecular surface $G(K)$. Depending on the model surface chosen, there will be other parameters in addition to the configuration K needed to specify $G(K)$ completely. In the case of electronic isodensity contours, one must specify the chosen density value. For VDWSs, it is the chosen atomic radii.

Once the model surface is chosen, we can associate with it a *discrete shape descriptor*, indicated by $\tau(G(K))$. Molecular topology provides a number of alternatives to derive three-dimensional shape descriptors [19–22]. The most convenient choice depends on the model representation for G. The main requirement is that $\tau(G(K))$ has to be discrete, i.e., that it can be expressed as a finite series of numbers. These numbers *do not necessarily change with small changes in the parameters specifying the surface (including the configuration K).* Molecular volumes and surface areas do not fit this criterion. In contrast, topological invariants of G (or of surfaces derived from G) are discrete.

Given a model surface and a shape descriptor, surfaces for different conformations can be classified according to their *shape type*. The shape type is indicated as $s(\tau,\ K)$. One can define an equivalence relation s between nuclear configurations in terms of molecular

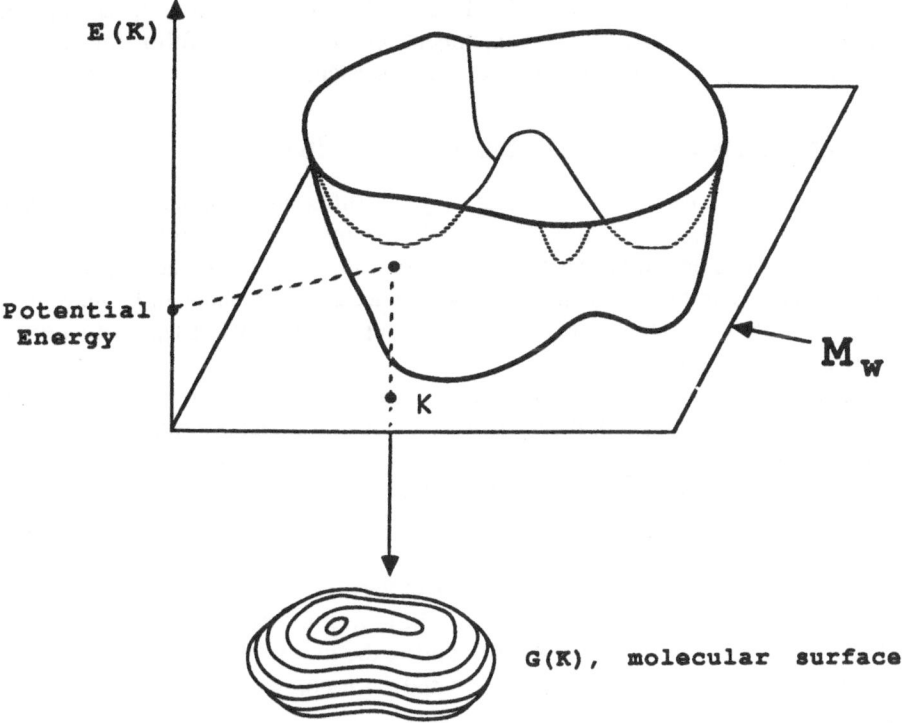

Fig. 1. Relation between energy, molecular shape, and nuclear configurational space. The drawing sketches a subset of the configurational space (M_w) where the energy is studied. One can associate a potential energy value $E(K)$ with every generic configuration K. As well, one can associate with K a molecular surface (or a set of surfaces) $G(K)$. A configurational rearrangement can be represented as a path on M_w. Along this path, the energy $E(K)$ and the surface $G(K)$ change. However, some of the shape features of $G(K)$ may remain invariant.

shape as follows:

$$K \; s(\tau) \; K', \quad \text{iff}: \quad s(\tau, \; K) = s(\tau, \; K'), \quad K, K' \in M \; . \tag{2}$$

The elements of the quotient set $M/s(\tau)$ are the equivalence classes of configurations characterized by having the same shape features on their associated molecular surfaces. These equivalence classes are the so-called *shape invariance domains* S_i (or "shape regions"). They provide a partitioning of the configuration space [23, 29], based on the values taken by the shape descriptor τ,

$$S_i(\tau) = \{K \in M : s(\tau, \; K) = s(\tau_i)\} \subset M \; , \tag{3a}$$

$$S_i \cap S_j = \vee, \quad i \neq j \; , \tag{3b}$$

$$M = \bigcup_i S_i \; , \tag{3c}$$

and another topological structure to MM. This topology is indicated as T_s, as opposed to the reaction topology T_C based on the catchment regions [10]. In a following section we compare these two partitionings for some chemical systems.

4. Shape of Flexible Molecules and Molecular Shape Dynamics

The recognition of similarities in molecular shape for different compounds is a central task towards the rational design of drugs [12, 32–37]. Compounds with targeted biochemical and pharmacological activities can be modeled on the assumption that they will exhibit properties comparable to those compounds which are closer in shape to them. Similarly, they can be modeled by taking into account the complementarity in shape with a receptor cavity with which they interact.

However, molecules are nonrigid objects. The interaction of a molecule with a receptor does not depend on the details of the nuclear locations or the details of the molecular shape but rather on *the average shape defined by the molecular flexibility*. Although a number of approaches have been proposed to assess molecular shape quantitatively, it is only recently that attention has been paid to the extent to which the results depend on the molecular flexibility [23–31].

Taking into account nonrigidity (in the form of various small and large amplitude oscillations), a purely geometrical approach to describe molecular shape becomes inconvenient. The actual nuclear positions change locally during the motion, whereas some global features may remain unchanged. Approaches based on topological rather than geometrical properties are more realistic [23–31].

Molecular flexibility can be described in different ways. The simplest method is to describe large and small amplitude vibrations semiclassically by means of molecular dynamics and statistical mechanics. This is virtually the only practical alternative in the case of macromolecules [38–40]. In the case of medium size or small molecules, oscillations can be computed within the context of quantum mechanics. In all cases, the information obtained corresponds to vibrational amplitudes and trajectories in 3-space.

As mentioned before, the quantitative characterization of shape depends on the model used for the molecular surfaces. The results reviewed in this work use a family of topological methods collectively known as *shape group methods*. The approach is different whether the molecular surface is described as a continuum of isodensity contour surfaces or of VDWSs. The techniques have been thoroughly discussed in the literature [19–31]. A detailed discussion falls outside the scope of this work. For what follows, one must only bear in mind that these techniques provide lists of numbers (discrete shape descriptors τ), which are related to the topological properties of the surfaces. These numbers are the descriptors we shall use to label the shape invariance regions in configurational space. These descriptors are shape codes which can be used to monitor the changes, if any, introduced by the different motions accessible to a molecule.

The description of molecular flexibility given above deals with the vibrations of an isolated species. A more realistic approach should take into account environmental effects such as the electric fields caused by other molecules and the solvent. Several articles have dealt with the effect of external fields on molecular structure (see [41–44]). Recently, topological methods of shape characterization have been employed to characterize electron densities distorted by weak electric fields [45]. The tools and techniques necessary to incorporate these external perturbations on reaction paths are basically the same ones that will be discussed in the next sections.

5. Changes of Molecular Shape along Reaction Paths: Shape Transitions

A reaction path in M is a one-dimensional subset of configurations K (i.e., a trajectory in M). Such a path can be represented parametrically as a continuous assignment of points from the unit interval $I = [0, 1]$ to configurations K. Formally, this can be expressed as a mapping between topological spaces:

$$p(K_0, t): \ (I, T) \rightarrow (M, T_M), \tag{4}$$

where $K_0 = p(K, 0)$ is the starting point of the path and T and T_M are the metric topologies of I and M, respectively. Many definitions of reaction paths have been given [46–49]. In this work we shall restrict the analysis to minimum energy paths based on Fukui's definition [49].

In previous works, we have discussed the changes in molecular shape along a parametrized path $p(K_0, t)$ [23–30]. In our case the shape of a molecule is described by a continuum of molecular sufaces. One continuum is assigned to each configuration K.

The energy varies continuously along path $p(K, t)$. This change is usually represented by reaction profiles in terms of a reaction coordinate. In most cases, we shall take the reaction coordinate as the arc length of the reaction path, defined with reference to mass-weighed cartesian coordinates [10].

In contrast with the energy, the essential shape features of a molecule (as defined by shape types) change discretely along $p(K, t)$. These changes are the so-called *shape transitions*. If the shape is described in terms of *a single* molecular surface $G(K)$, $K = p(K_0, t)$, then there is a *unique* value of the parameter t identifying the nuclear configuration where a new shape type occurs. On the other hand, *a continuum* of surfaces will exhibit a shape transition over a section of the path. This section of the path plays the role of a *boundary range* between the two consecutive shape types. The location and width of these boundary ranges can be used to assess degrees of molecular similarity along the path.

As a first illustration, we analyze a simple [1, 3] hydrogen-transfer reaction. The system considered is the formic acid-formic acid "isomerization". The details on the computation of the reaction path, molecular shapes, and their interrelations are given in [29]. Below we discuss only the essential results.

Figure 2 compares the effect of two very different types of configurational rearrangements on the electron density. The upper diagrams represent the atomic nuclei at their equilibrium positions, with the arrows pointing their motions away from equilibrium. The molecule remains planar during these distortions.

The effect on the molecular shape is depicted by the electron isodensity cross section at the molecular plane. The figure shows the deformation of the contour line corresponding to the electron density value of 0.002 a.u. This value provides a reasonable representation of molecular size and shape [50–52].

The scheme on the left-hand side of Figure 2 represents an almost pure O—H stretching. Note that the electron density is deformed but *the overall curvature features* are preserved. This demonstrates that, though the nuclear geometry varies, the molecule remains essentially unchanged by pure bond vibration.

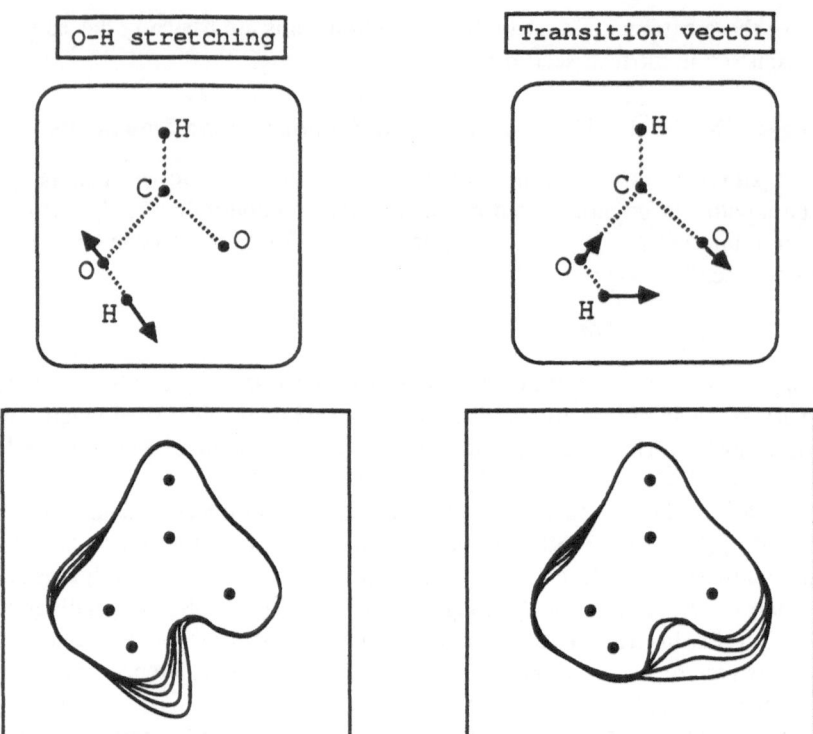

Fig. 2. Effect on the electron density cross sections of formic acid of various configurational rearrangements. Each contour line corresponds to a sketch of the cross section for the electron density value of 0.002 a.u. The figure on the left-hand side represents an O—H stretching. The electron density is deformed but the overall curvature features are preserved. In contrast, the figure on the right-hand side shows that the shape of the contour is affected more strongly by the deformation leading to the transition structure. Black dots are used for the equilibrium geometry.

The right-hand side figures show the changes in the contour lines during the deformation leading to the transition structure. This is the typical distortion found when leaving the equilibrium geometry along the isomerization reaction path. The contrast is noticeable. The electron density is strongly distorted by a motion of the nuclei to the same extent as that considered before for the vibration. Note that both the curvature and symmetry of the electron density contour change. These results indicate that the shape of a molecular surface depends strongly on the type of configurational rearrangement considered.

Figure 3 shows the overall result for this reaction by providing the partitionings of a restricted configurational space M_w. The working subset M_w is defined by the distances from the migrating hydrogen atom to the two oxygen atoms, that is, by two O—H stretchings. The upper map corresponds to the partitioning of M_w in terms of the shapes of the molecular surfaces. Each shaded or labelled region represents a different shape type $s(\tau, K)$. For simplicity, we have used the results from [29] for the shapes of the corresponding VDWSs instead of using density contours. Van der Waals surfaces with the standard radii provide a reasonable approximation to the 0.0025 a.u. isodensity surfaces [52]. In our approach, we have used a set of fused sphere surfaces defined by a continuous variation of the atomic radii, accounting for a range of electron densities of 0.0025 ± 0.0005 a.u. [52]. All nuclear geometries and electronic energies in the diagram have been computed at *ab initio* RHF level using the basis set 3–21G with the program Gaussian 88 [53].

The label R identifies all configurations that, within the range of atomic radii, give rise to a VDWS whose shape descriptors coincide with a descriptor associated with the reactant's configuration. TS corresponds to the equivalent situation for the transition structure. Due to the symmetry, we have not made a distinction between reactants and products. Shape regions labeled with numbers correspond to shape types other than those associated with the critical points. Note that these additional shapes occur for large values of both O—H stretchings. In this region, the VDWS (or electron isodensity contour) shows the separation of a hydrogen atom from the total molecular surface.

Note that the shape region for the reactants (R) is more extended along one O—H direction. This is in agreement with the previous observation that the molecular shape is preserved during a rather large O—H vibration. In contrast, the shape region for TS is extended in a direction involving a simultaneous change in both O—H distances, as it would be expected from the transition vector.

The lower map in Figure 3 corresponds to the partitioning obtained by using the potential energy function. The regions R and TS are the catchment regions associated with the energy minima and saddle point, respectively.

The minimum energy reaction path is indicated by the heavy line. In the energy partitioning, all configurations of the path belong to the catchment region R, except the saddle point itself. Consequently, an infinitesimal deformation along the path leads to a different "chemical species." In contrast, the shape partitioning shows that it is necessary to move along $p(K, t)$ far from TS in order to distort sufficiently its electron density (or VDWS) and change its shape into that of R.

This reaction illustrates the type of insight gained by analyzing not only energy but also molecular shape along a reaction path. The partitioning based on shape types expresses quantitatively the physical notion that a small oscillation about a saddle point leads to configurations similar to the TS but not necessarily to the reactants and products.

The information in Figure 3 can be rendered in a more familiar form by representing the changes of energy and molecular shape in a reaction profile. Figure 4 presents the reaction barrier for the isomerization. Here, the reaction coordinate is given by the integrated arc length of the reaction path computed in a mass-weighed cartesian frame (with units of $\mathring{A}$ (a.m.u.)$^{1/2}$).

The shaded regions indicate the extent of the reaction path associated with each of the three shape types encountered in Figure 3. It is immediately apparent up to which degree the formic acid must be deformed in order to arrive at a configuration which resembles the TS.

This analysis of energy profiles provides useful information on a chemical reaction. When studying a section of the path involving two consecutive critical points, it is often found that there is a range of conformations whose shapes are those of either critical point depending on the atomic radii or electron density values chosen. The position and width of this boundary range are related to the energy difference between critical points. The consequences of these relationships on molecular similarity are commented on in the next section.

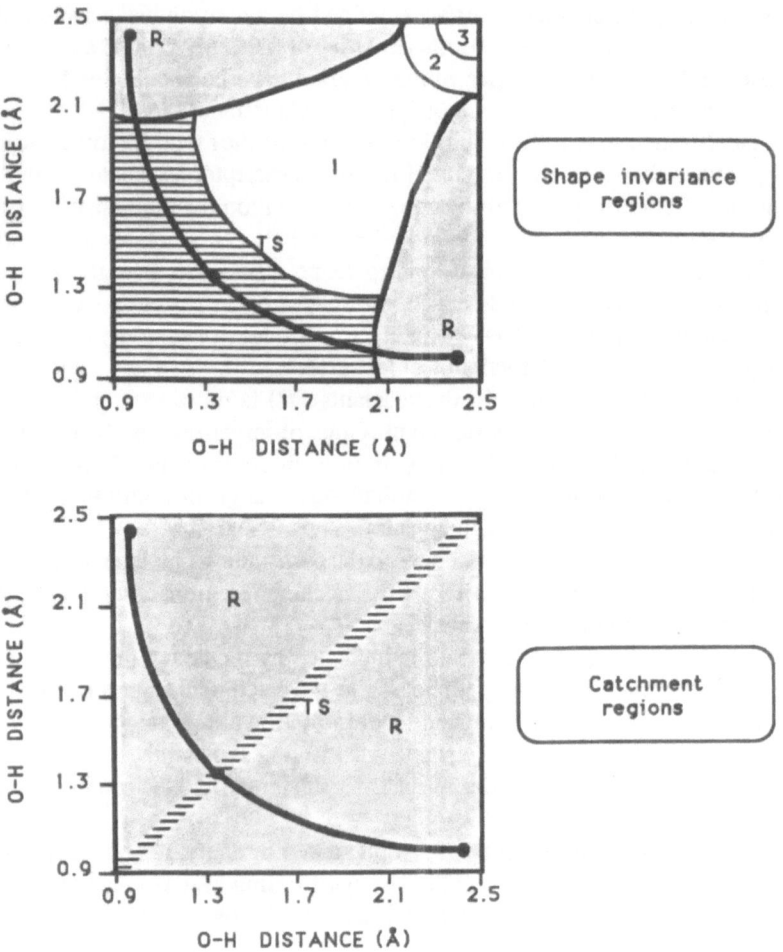

Fig. 3. Comparison between the shape invariance and catchment region partitionings of a 2D configurational subset M_w for the [1, 3] hydrogen-shift isomerization of formic acid.

To conclude, we discuss another isomerization reaction. Figure 5 shows the potential
energy function for the keto-enol isomerization from vinyl alcohol to acetaldehyde. The
restricted space M_w is defined by the C—C distance and C-H distance to the migrating
hydrogen atom. These two distances are "hard coordinates", but their change characterizes
the reaction that involves the transformation from a double to single CC bond and the
formation of a new CH bond. All other internal coordinates have been optimized at a
semiempirical level using Dewar's MNDO method [54, and others quoted therein] as
implemented in Gaussian 88 [53].

Figure 6 gives the partitioning of M_w into shape regions. A comparison with Figure 3
reveals similar features. For instance, the C—C vibration does not affect much the shape of
the molecular surface. A change in shape requires simultaneous C—C and C—H distortions,

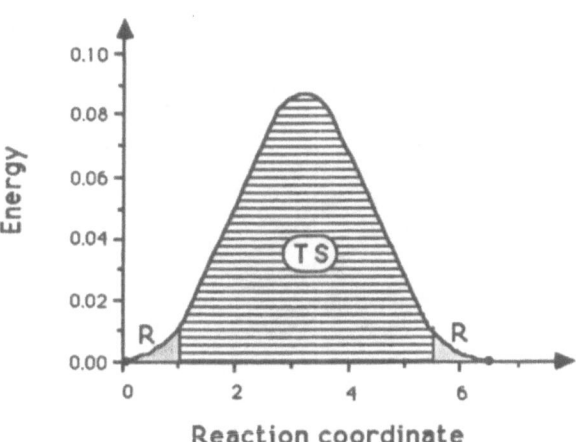

Fig. 4. Reaction barrier for the [1,3] hydrogen-shift isomerization of formic acid. The minimum energy reaction path has been computed at the SCF 3–21G *ab initio* level. Energy is measured in atomic units. The reaction coordinate is the integrated arc length of the path (the units are $\overset{\circ}{A}$ (a.m.u.)$^{1/2}$). The various shape invariance domains encountered along the path are indicated by shaded regions. R and TS correspond to regions of configurational space where the molecular shape is the same as that of the reactant and transition structure, respectively.

e.g., a motion in the direction of the transition vector.

6. Consequences: Molecular Similarity and Shape-energy Rules

The occurrence of one or more shape types along a reaction path provides structural information that may be hidden when analyzing only the potential energy. Consider, for example, the case of a path connecting two minima through a transition structure. As seen in the previous section, the configurations along the reaction path can be classified only as belonging to three catchment regions. This fact is not affected by a change in the barrier's height as long as there are three critical points. However, it is conventionally assumed that the smaller the barrier, the more similar the structures of the connected critical points (Hammond-Loeffler postulate [55–57]). This relationship between energy differences and degrees of similarity in molecular shape can be given a quantitative expression by using the approach discussed previously.

Let us introduce some parameters describing the shape changes along a section of a reaction path. Figure 7 presents a scheme of a generic barrier connecting three critical points ($R \rightarrow TS \rightarrow P$).

The figure takes into account the frequent case of a *boundary range* between the shapes of two consecutive critical points. This boundary range is defined as the set of configurations whose molecular surfaces have the shape type of the neighboring critical points depending

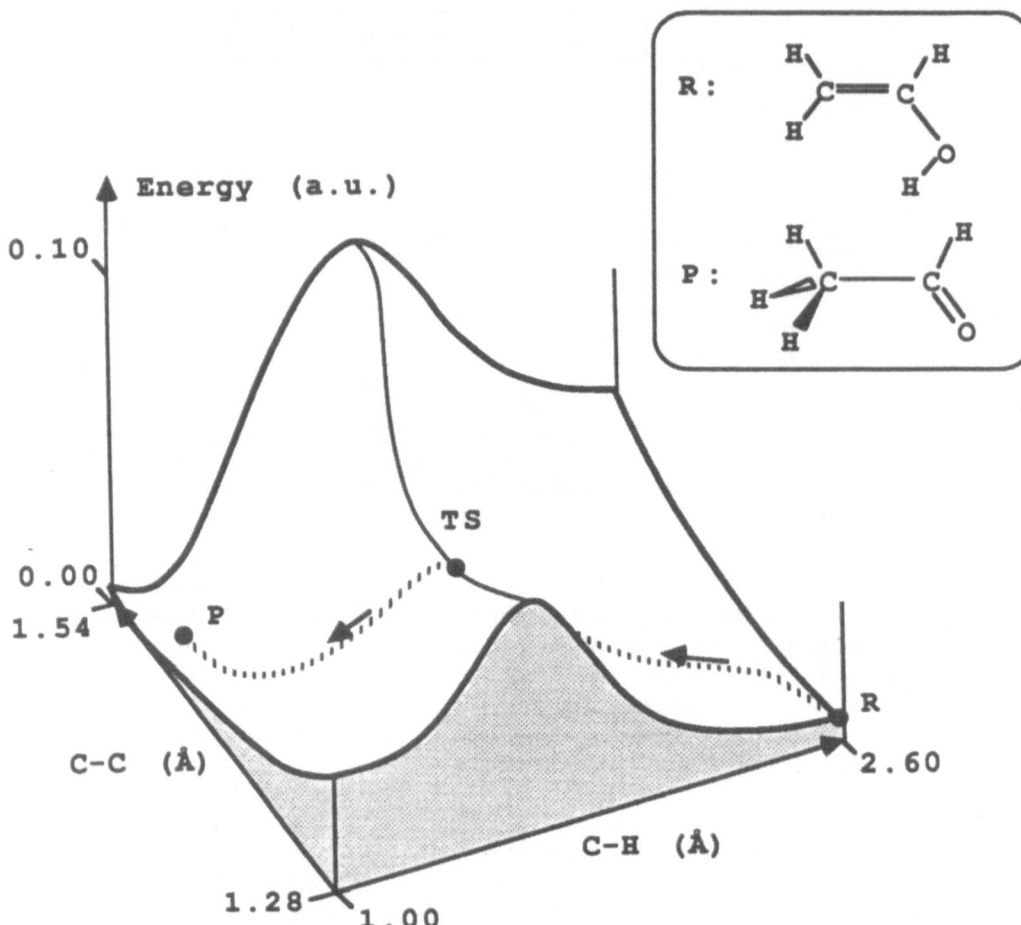

Fig. 5. Potential energy surface for the isomerization of vinyl alcohol to acetaldehyde restricted to a 2D subset M_w. The reactant (R) and product (P) are indicated in the upper right corner. The energies are computed at MNDO level with full geometry optimization outside M_w. Black dots indicate the critical points and the dashed curve is the reaction path leading from R to P.

on the values given to the parameters specifying the surfaces (electron density values or atomic radii). This intermediate region represents an *uncertainty in the shapes and it accounts for the lack of a precise definition of the boundary of a molecule.*

The uncertainty (width) of the boundary range between shape invariance regions often correlates with the energy difference between consecutive critical points. Generally, the smaller the energy difference, the wider the boundary range. Degrees of similarity can be assessed in terms of the width of this intermediate region.

There are some differences between the changes in shape and the changes in symmetry along reaction paths. We summarize some of the findings [27–30] which are important for our discussion later:

i) Whereas consecutive segments of the path not containing critical points conserve the

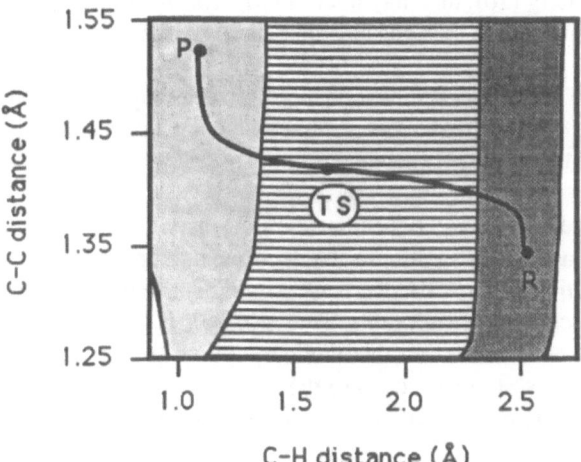

Fig. 6. Shape invariance map for the keto-enol isomerization from vinyl alcohol (R) to acetaldehyde (P) computed at MNDO level. The working configurational subset M_w is defined by the C—C distance and C—H to the migrating hydrogen atom. The CC stretching illustrates the transformation from the doubly-bonded vinyl alcohol (C—C distance 1.35 Å) to the singly-bonded acetaldehyde (C—C distance 1.52 Å). The C—H coordinate describes the migration of a hydrogen atom from being bonded to the oxygen atom (C—H distance of 2.53 Å) to becoming bonded to a carbon atom (C-H distance of 1.06 Å). The minimum energy (Fukui) path is indicated by the heavy line.

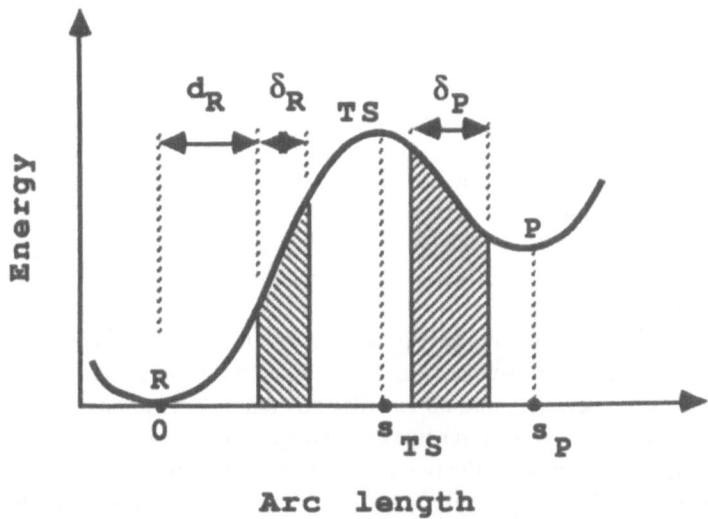

Fig. 7. Generic double-well reaction barrier. There are two stable species R and P (reactant and product) and a transition structure (TS). The change in potential energy is followed by considering the reaction path arc length the reaction coordinate. Shaded regions represent the shape boundary ranges found along the path (i.e., the configurations whose molecular shape is the same as those of the neighboring critical points, depending on the molecular surface chosen). The widths δ_R and δ_P and the distance d_R are normalized to $s_{TS} = 1$. The distance d_R represents the fraction of the reaction path separating the reactant from the first boundary range.

point group symmetry [10], this may not hold true for the molecular shape.

ii) Critical points at the extremes of the path with different point groups can have the same shape type.

iii) There is no one-to-one correspondence between shape invariance regions and catchment regions. It is possible that every critical point is associated with a distinct shape type, but there are usually a number of shape regions not associated with any critical point (cf. Figures 3 and 6).

We have studied the energy dependence of the shape parameters indicated in Figure 7: a) distance d, measuring the path arc length from one critical point up to the boundary range. For convenience, this distance is measured as a fraction of the section of the path joining two consecutive critical points. Thus, $d_R \leq 1$ is the distance from the reactant to the boundary range; b) width δ of the boundary region (as a fraction of the section of the path). The width $\delta_R \leq 1$ corresponds to the $R \to TS$ transformation. Note that $d_R \to 0$ if $\delta_R \to 1$, i.e., the two consecutive critical points cannot be distinguished according to shape.

In [27], the dependence of d_R and δ_R on the energy change ΔE was studied for a number of reactions. Here, we discuss the results for a series of simple three-body reactions. These are the collisional processes leading to the formation of HNO and NO_2 in various electronic states [27]:

$$H(^2S) + NO(^2\Pi) \to HNO(^1A', \, ^3A'', \, ^1A''),$$ (5a)

$$N(^4S) + O_2(^2\Pi) \to NO_2(^2A') \to O(^3S) + NO(^2\Pi),$$ (5b)

as well as the isomerizations [27]:

$$HCN \to CNH,$$ (5c)

$$BCN \to CNB.$$ (5d)

Figure 8 displays the dependence of d_R on the potential energy change for the different path sections. Note that d_R properly refers to the reactant in the R→TS transformation, but it corresponds to the TS in the second half of the path (TS→P). All energies are high-level *ab initio* results [27]. The shapes are characterized by the shape group method for continua of VDWSs [22].

The correlation found provides a quantitative expression to the notion that the structural similarity between consecutive critical points along a path decreases with an increment in their energy differences (Hammond-Loeffler postulate). An increment of d_R expresses the fact that a change in shape features involves a big distortion in the geometry if ΔE is large. This deformation is smaller if the two critical points are closer in energy.

Figure 9 complements the discussion by showing the change of the width of the boundary range with the energy difference for the same set of reactions. The results show an opposite behavior to that in Figure 8. That is, the smaller the energy difference the wider the boundary range, implying that *the two consecutive critical points are not distinguishable in shape over a large section of the path*. In contrast, *the shape transition becomes sharper with an increase of the energy ΔE*.

This tendency is in agreement with the observations in Figure 8. Note that a large degree of structural similarity between two critical points implies that:

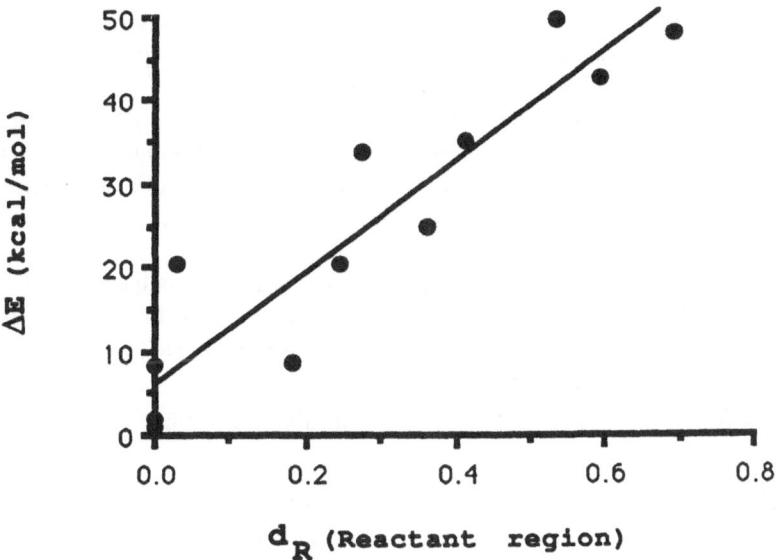

Fig. 8. Relation between the energy change ΔE and the fraction d_R of a path associated with the transformation from the shape of the starting critical point to the shape of the boundary range. The sections of paths considered are for the reactions (5). Data are from [27].

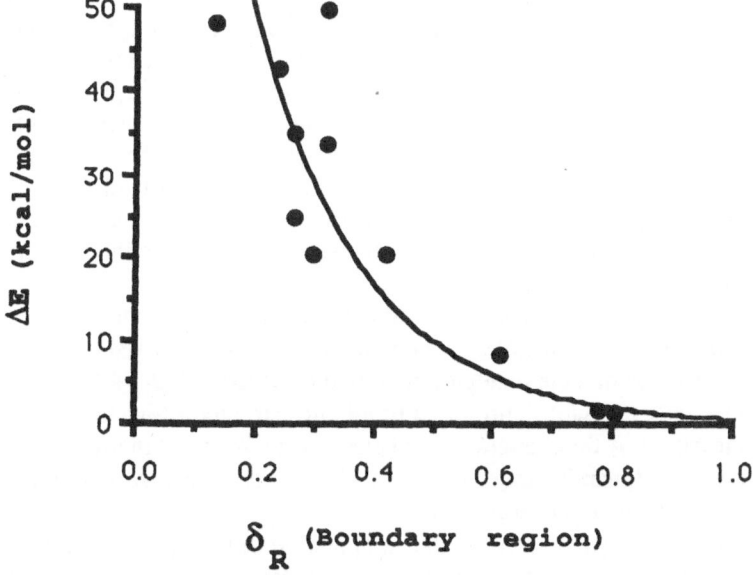

Fig. 9. Relation between the energy change ΔE and the fraction δ_R of the path corresponding to the width of the boundary regions between the reactant and the TS shapes. The results correspond to sections connecting critical points in the reaction paths for the reactions (5).

i) a small deformation brings the molecular shape type associated with one critical point to the shape types associated with configurations in the boundary region;

ii) the boundary region is wide, since the two neighboring shape types can easily be interconverted by a small change in the molecular surface.

For the reactions analyzed here, the trend between ΔE and d_R shows a cut-off value of about 6 kcal/mol. Below this value, two consecutive critical points appear to be undistinguishable in shape. The validity of these trends for more complicated conformational rearrangements requires further study. Some additional results for ring-opening reactions appear in [58].

In summary, the parameters d and d are related to the boundary range wherein a molecule is described as having more than one shape type. The results show that these simple barrier parameters can provide measurements of structural similarity [27].

7. Shapes of Macromolecules and Computer-simulated Folding

For large biomolecules, it is necessary to follow a different approach to analyzing molecular shapes along reaction paths. In these cases, the use of continua of van der Waals surfaces or density contours is not only impractical, but more importantly, the continua do not depict the features most relevant to the understanding of biomolecular properties. This is especially true when one is interested only in recognizing details of the structural organization of proteins and monitoring their change along folding paths.

Many properties of proteins depend on general characteristics of their fold [59, 60]. Richardson [61] has provided a classification of some of the basic folding patterns. Their role in determining properties and functions of proteins are becoming apparent (see, for instance, [62–64]). However, the recognition of these motifs is not a trivial task. Most of the techniques used so far use local features of sections of the backbone, such as dihedral angles (Ramachandran diagrams [59, 60]) and distances between segments [65]. Yet, local features appear to be more relevant when analyzing properties of proteins other than their fold (e.g., the recognition and matching of active sites [66]). Global descriptions of shape such as those provided by molecular topology should be used instead.

The simplest representation of a protein fold is given by its backbone. This molecular skeleton is defined by the sequence of a-carbons, disregarding the internal structure of the side chains of the amino acid residues. A number of techniques have been proposed in the literature to characterize the backbones. Simple topological invariants such as writhing and linking numbers [67–71] have been used. Other procedures involved the evaluation of the curvature and torsion of a curve approximating the backbone [72–74].

There are other techniques proposed to monitor structural transitions in proteins. For example, one can study the changes in local principal moments of inertia of a helical section along a molecular dynamics trajectory [75]. These approaches are not global and do not provide discrete shape descriptors.

Topological techniques based on the computation of knot polynomials and molecular graphs derived from backbone overcrossings have also been used for detailed shape characterizations [76–79]. The crossing pattern of sections of the backbone provides essential topological information on the space curve. This information can be stored in the form of vectors with dimensionality equal to the number of crossings and crossing indices (i.e., handednesses) as elements [76–79].

Recently, this procedure has been applied to the characterization of small proteins [77, 80, 81]. The methodology takes into account the overcrossing of backbone segments when viewing the molecule along a given direction in space. The crossing pattern characterizes the view. In considering all possible views in 3-space, we construct *a spherical shape map of topological invariants describing some essential features of the protein's shape* [80].

This method has been employed to study the structural and shape stability in molecular dynamics simulations of protein unfolding paths [81]. The procedure serves as a useful tool for recognizing different degrees of organization in a protein fold and for making quantitative assessments on the persistence of the various secondary structural features throughout the folding process. For example, the effect of small amplitude vibrations are distinguished from those leading to partial or complete unfolding of helical loops. These two rearrangements lead to quite different crossing patterns [81].

8. Future Directions

The methodologies reviewed here are new tools whose potential is becoming evident only now. One can foresee various directions for further development and applications. As a conclusion of this work, we describe some of the problems that can be addressed:

i) Full three-dimensional molecular shape characterization and modeling of small bio-organic molecules. As discussed above, one can have a very detailed description of molecular shape which includes deformation of the electron density due not only to the intrinsic molecular flexibility but also to environmental effects. Properties other than the electron density, e.g. the electrostatic potential, can be also brought into the picture. This approach can have numerous applications in the study of active sites by computer-assisted receptor mapping.

ii) Molecular shape and dynamical changes of proteins. The extension of these methodologies to their application to proteins provides a method useful in recognizing the occurrence and formation of structural motifs. Moreover, it gives a structural and shape classification for irregular proteins where it may not always be possible to recognize any standard supersecondary organization. As above, one can take into account environmental effects on the molecular shape.

iii) Design of dynamical models of drug-receptor complexes (including conformational flexibility and thermal motion). Elements of the above two tasks are combined here for the analysis of drug molecules together with protein receptors. The analysis of shape complementarity between interacting molecules, as well as for molecular recognition, can be addressed in this context.

Acknowledgments

The author would like to thank Ms. Naomi D. Grant for her collaboration in the computation of electron densities for the isomerization of formic acid. As well, the support of NSERC (Canada) to the Topology Program under the direction of P.G. Mezey, in the form of both operating and strategic grants, is gratefully acknowledged.

References

[1] F.M. Richards: *Annu. Rev. Biophys. Bioeng.* 6 (1977) 151.

[2] R. Langridge, T.E. Ferrin, I.D. Kuntz, and M.L. Connolly: *Science* **211** (1981) 661.
[3] P.A. Basch, N. Pattabiraman, C. Huang, T.E. Ferrin, and R. Langridge: *Science* **222** (1983) 1325.
[4] M.L. Connolly: *J. Am. Chem. Soc.* **107** (1985) 1118.
[5] R.S. Pearlman: in *Partition Coefficient: Determination and Estimation*, W.J. Dunn, J.H. Brock, and R.S. Pearlman, Eds., Pergamon, New York, 1986.
[6] J.C. McGowan and A. Mellors: *Molecular Volumes in Chemistry and Biology: Applications Including Partitioning and Toxicology*, Horwood, Chichester, 1986.
[7] A.Y. Meyer: *Chem. Soc. Rev.* **15** (1986) 449.
[8] H.R. Karfunkel and V. Eyraud: *J. Comput. Chem.* **10** (1989) 628.
[9] L. Schäfer: *J. Mol. Struct.* **100** (1983) 51.
[10] P.G. Mezey: *Potential Energy Hypersurfaces*, Elsevier, Amsterdam, 1987.
[11] R.F.W. Bader: *Atoms in Molecules*, Oxford U.P., Oxford, 1990.
[12] M.A. Johnson and G.M. Maggiora (eds.): *Concepts and Applications of Molecular Similarity*, Wiley, New York, 1990.
[13] P.G. Mezey: in *Reviews in Computational Chemistry*, K.B. Lipkowitz and D.B. Boyd (eds.), VCH, New York, 1990.
[14] a) L. Lathouwers and P. Van Leuven: *Adv. Chem. Phys.* **49** (1982) 115 (and previous refs. quoted therein).
     b) L. Lathouwers, P. Van Leuven, E. Deumens, and Y. Öhm: *J. Chem. Phys.* **86** (1987) 6352.
[15] H. Primas: in *Quantum Dynamics of Molecules: the New Experimental Challenge to Theorists*, Ed. R.G. Woolley, NATO ASI B 57, Plenum Press, New York, 1980.
[16] R.G. Woolley: *Struct. Bonding* **52** (1982) 1; *Chem. Phys. Lett.* **125** (1986) 200.
[17] P. Claverie: in *Symmetries and Properties of Non-Rigid Molecules*, J. Maruani and J. Serré (eds.), Elsevier, Amsterdam, 1983.
[18] H.J. Monkhorst: in *Modelling of Structure and Properties of Molecules*, Z.B. Maksic (ed.), Ellis Horwood, Chichester, 1987.
[19] P.G. Mezey: *Int. J. Quantum Chem. QBS* **12** (1986) 113.
[20] P.G. Mezey: *J. Comput. Chem.* **8** (1987) 462.
[21] P.G. Mezey: *J. Math. Chem.* **2** (1988) 299.
[22] G.A. Arteca and P.G. Mezey: *J. Comput. Chem.* **9** (1988) 554.
[23] G.A. Arteca and P.G. Mezey: *Int. J. Quantum Chem. QBS* **14** (1987) 133; ibid. *QBS* **15** (1988) 33.
[24] G.A. Arteca and P.G. Mezey: *Int. J. Quantum Chem. QCS* **23** (1989) 305.
[25] P.G. Mezey: *J. Math. Chem.* **2** (1988) 325.
[26] G.A. Arteca and P.G. Mezey: *Int. J. Quantum Chem. QCS* **24** (1990) 1.
[27] G.A. Arteca and P.G. Mezey: *J. Phys. Chem.* **93** (1989) 4746.
[28] G.A. Arteca and P.G. Mezey: *Int. J. Quantum Chem.* **38** (1990) 713.
[29] G.A. Arteca and P.G. Mezey: *J. Mol. Struct. (Theochem)* **230** (1991) 323.
[30] P.W. Walker, G.A. Arteca, and P.G. Mezey: *J. Comput. Chem.* **12** (1991) 220.
[31] W.G. Richards: *Quantum Pharmacology*, Butterworths, London, 1983.
[32] I. Motoc: 'Steric and Other Structural Parameters for QSAR', in A.T. Balaban, A. Chiriac, I. Motoc, and Z. Simon (eds.), *Steric Fit in Quantitative Structure-Activity Relationships*, Lecture Notes in Chemistry No. 15, Springer Verlag, Berlin, 1980.
[33] A.J. Stuper, W.E. Brügger, and P.C. Jurs: *Computer-Assisted Studies of Chemical Structure and Biological Function*, Chap. 3, Wiley, New York, 1979.
[34] J.M. Blaney, E.C. Jorgensen, M.L. Connolly, T.E. Ferrin, R. Langridge, S.T. Oatley, J.M. Burridge, and C.C.F. Blake: *J. Med. Chem.* **25** (1982) 785.
[35] R. Franke: 'Theoretical Drug Design Methods', Chap. 12, *Pharmacochemistry Library*, Vol. 7, Elsevier, Amsterdam, 1977.
[36] J.C. Dearden (ed.): 'Quantitative Approaches to Drug Design', *Pharmacochemistry Library*, Vol. 6, Elsevier, Amsterdam, 1983.
[37] M. Karplus and A. McCammon: *CRC Crit. Rev. Biochem.* **9** (1981) 292.
[38] A. McCammon and S.C. Harvey: *Dynamics of Proteins and Nucleic Acids*, Cambridge, London, 1987.
[39] O. Tapia: in *Molecular Interactions*, H. Ratajczack and W.J. Orville-Thomas (eds.), Vol. 3, Wiley, New York, 1982.
[40] G.B. Bacskay and N.S. Hush: *Theor. Chim. Acta* **32** (1974) 311.
[41] M. Raynaud, C. Reynaud, Y. Ellinger, G. Henrico, and J. Delhalle: *Chem. Phys.* **142** (1990) 191.
[42] M.P.C.M. Krijn and D. Feil: *J. Phys. Chem.* **91** (1987) 540.
[43] M. Solà, A. Lledós, M. Duran, and J. Bertrán: *Int. J. Quantum Chem.* **40** (1991) 511.
[44] G.A. Arteca and P.G. Mezey: *J. Mol. Struct. (Theochem)*, in press; *Chem. Phys.*, to be published.

[45] H. Eyring and M. Polanyi: Z. Phys. Chem. B12 (1931) 279.
[46] K. Fukui: J. Phys. Chem. 74 (1970) 4161.
[47] P. Pechukas: J. Chem. Phys. 64 (1976) 1516.
[48] A. Tachibana and K. Fukui: Theor. Chim. Acta 49 (1978) 312; ibid. 51 (1979) 189.
[49] R.F. Hout and W.J. Hehre: J. Am. Chem. Soc. 105 (1983) 3728.
[50] M.M. Francl, R.F. Hout, and W.J. Hehre: J. Am. Chem. Soc. 106 (1984) 563.
[51] G.A. Arteca, N.D. Grant, and P.G. Mezey: J. Comput. Chem., in press.
[52] M.J. Frisch, J.S. Binkley, H.B. Schlegel, K. Raghavachari, C.F. Melius, R.L. Martin, J.J.P. Stewart, F.W. Bobrowicz, C.M. Rohlfing, L.R. Kahn, D.J. Defrees, R. Seeger, R.A. Whiteside, D.J. Fox, E.M. Fleuder, and J.A. Pople: Gaussian 88, Carnegie-Mellon Quantum Chemistry Publishing Unit, Pittsburgh, 1988.
[53] M.J.S. Dewar, E.G. Zoebisch, E.F. Healy, and J.J.P. Stewart: J. Am. Chem. Soc. 107 (1985) 3902.
[54] G.S. Hammond: J. Am. Chem. Soc. 77 (1955) 335.
[55] L. Melander: The Transition State, Royal Chemical Soc. Special Publications, London, 1962.
[56] G.A. Arteca and P.G. Mezey: J. Comput. Chem. 9 (1988) 728.
[57] X. Luo, G.A. Arteca, and P.G. Mezey: Int. J. Quantum Chem., in press.
[58] C.R. Cantor and P.R. Schimmel: 'The Conformation of Biological Macromolecules', Biophysical Chemistry, Part I, Freeman, San Francisco, 1980.
[59] G.E. Schulz and R.H. Schirmer: Principles of Protein Structure, Springer-Verlag, New York, 1979.
[60] a) J.S. Richardson: Adv. Protein Chem. 34 (1981) 167.
b) J.S. Richardson: Methods in Enzymol. 115 (1985) 359.
[61] R. Jaenicke: Prog. Biophys. Molec. Biol. 49 (1987) 117.
[62] C. Chothia: Annu. Rev. Biochem. 53 (1984) 537.
[63] G.M. Maggiora, B. Mao, K.C. Chou, and S.L. Narasimhan: in Protein Structure Determination. Methods of Biochemical Analysis, C.H. Suelter (ed.), Wiley, New York, 1990.
[64] N.S. Goel and R.L. Thompson: Computer Simulations of Self-Organization in Biological Systems, Macmillan, New York, 1988.
[65] J. Åqvist and O. Tapia: J. Mol. Graph. 5 (1987) 30.
[66] R.B. Fuller: Proc. Natl. Acad. Sci. USA 68 (1971) 815 (and previous refs. quoted therein).
[67] T. Kikuchi, G. Némethy, and H.A. Scheraga: J. Comput. Chem. 6 (1986) 67.
[68] M. Le Bret: Biopolymers 18 (1979) 1709.
[69] P. De Santis, S. Morosetti, and A. Palleschi: Biopolymers 22 (1983) 37.
[70] M.-H. Hao and W.K. Olson: Biopolymers 28 (1989) 873.
[71] A.H. Louie and R.L. Somorjai: J. Theor. Biol. 98 (1982) 189.
[72] A.H. Louie and R.L. Somorjai: J. Mol. Biol. 168 (1983) 146.
[73] A.H. Louie and R.L. Somorjai: Bull. Math. Biol. 46 (1984) 745.
[74] O. Nilsson and O. Tapia: Proteins, to be published (& private communication).
[75] G.A. Arteca and P.G. Mezey: J. Mol. Graph. 8 (1990) 66.
[76] G.A. Arteca, O. Tapia, and P.G. Mezey: J. Mol. Graph. 9 (1991) 148.
[77] G.A. Arteca and P.G. Mezey: in Theoretical and Computational Models for Organic Chemistry, S.J. Formosinho, I.G. Csizmadia, and L.G. Arnau (eds.), NATO ASI, Kluwer, Dordrecht, 1991.
[78] G.A. Arteca and P.G. Mezey: in Structure, Reactivity, and Interactions, S. Fraga (ed.), Elsevier, Amsterdam, in press.
[79] G.A. Arteca and P.G. Mezey: Biopolymers, submitted.
[80] G.A. Arteca, O. Nilsson, and O. Tapia: J. Mol. Graph., submitted.

STRUCTURE OF MOLECULES: EXPERIMENTS AND THEORY

V.S. MASTRYUKOC
Department of Chemistry
University of Moscow
Moscow 117234
U.S.S.R.

and

L.A. MONTERO and J.R. ALVAREZ-IDABOY
Facultad de Química
Universidad de La Habana
Habana 10400
Cuba

ABSTRACT. Three relevant examples where experimental chemistry is closely linked to theory are described in the field of molecular structure studies. Ring inversion potentials, geometry of saturated cycles, and relationships between pure electronic properties (including geometrical relationships, like ring internal angles, and ionization potentials) are examined. Case studies are compared using both experimental methods, like electron diffraction, and either empirical or purely theoretical procedures. A close interdependence between theoretical and experimental results is demonstrated.

1. Introduction

This article consists of three relatively independent parts which are brought together under the same rather broad title as examples of interesting interactions between theory and experiment in structural chemistry. Several selected cases discussed below are meant to show the successes and shortcomings in both experiment and theory.

In structural chemistry, experimental methods were derived earlier than reliable computational or theoretical models. Depending on the method chosen, uncertainties in experimental determinations are considered as normal measurement errors. However, neither the molecular mechanical, nor semiempirical SCF, nor *ab initio* SCF theoretical models of molecular geometries are as accurate (to varying degrees) as the experiment values. And, this is not always dependent on the quality or complexity of the theoretical foundation which supports the procedure. In this case, inaccuracies with regards to test experimental values are interpreted differently; alternately either neglecting or confirming the quality of the theoretical procedure.

Is it possible to take a model molecular geometry, which is the minimum of the hypersurface of the theoretical energy in respect to spatial coordinates of nuclei, as the true

L. A. Montero and Y. G. Smeyers (eds.), Trends in Applied Theoretical Chemistry, 135–147, 1992.
© 1992 *Kluwer Academic Publishers.*

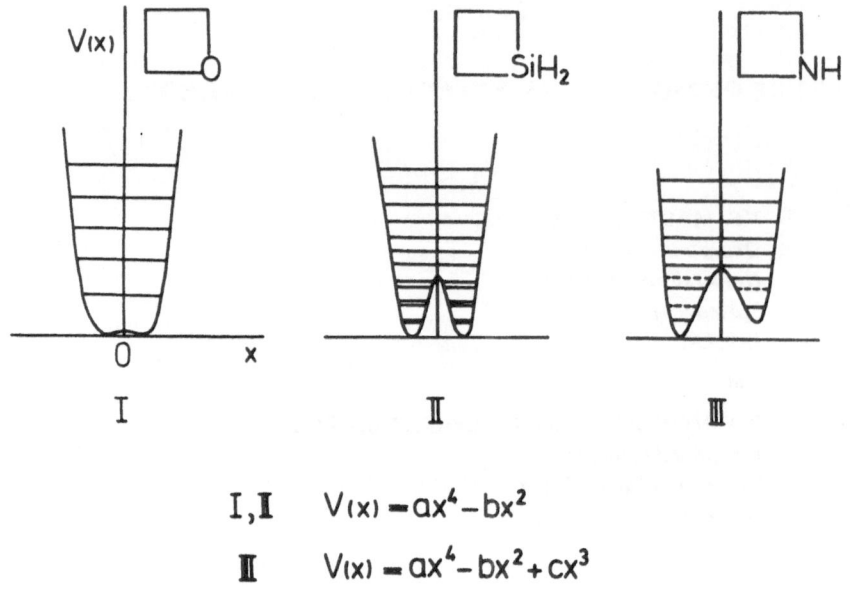

Fig. 1. Typical examples of one dimensional potential functions $V(x)$, where x is the ring puckering coordinate.

equilibrium geometry of the molecule, instead of the experimental data? How reliable is the comparison between experimental and theoretical geometrical values? Can theoretical models contribute to the understanding of significant details of molecular structure when the experimental data is not available or nonexistent? To establish criteria for these questions is the main purpose of the examples considered in this paper.

2. Asymmetrical Potential Functions for Ring Inversion

One dimensional potential functions, $V(x)$, for the ring inversion vibrations of a number of four- and five-membered rings have been calculated from vibrational transitions observed in the far infrared [1]. Some typical examples are shown in Figure 1.

Asymmetrical potential functions can have two non-equivalent minima and are called double minimum (DM) functions. Sometimes, however, the second minimum becomes a shoulder and results in a single minimum (SM) function. Both SM and DM possibilities for $V(x)$ are shown systematically in Figure 2.

In some cases it proved difficult to discriminate between these alternatives, especially when the second minimum was rather shallow (see more about it in [2]).

As far as the SM/DM controversy is concerned, we noted that two molecules, azetidine and bicyclo [3.1.0] hexane (BCH), seem to be each others opposite. Initially, DM and SM functions were reported for azetidine [3] and BCH [4] respectively. Later, the experimental data for the former were reinterpreted in favour of an SM function [5] and there is a growing

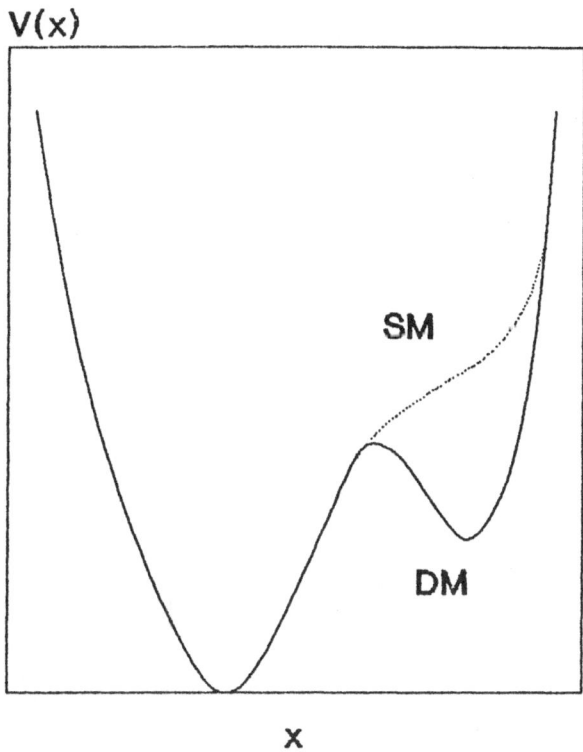

Fig. 2. One dimensional potential energy function $V(x) = ax^4 - bx^2 + cx^3$ with single minimum (SM) and non-equivalent minima (DM).

evidence that the exact opposite reinterpretation is required for BCH.

Azetidine: DM (1969) → SM (1981)

BCH: SM (1973) → DM (1990?) .

The key role of SCF *ab initio* calculations in both cases should be noted.

Azetidine: Two unequivalent minima on the DM function were rationalized in terms of coexisting equatorial **1** and axial **2** conformers [3].

On the other hand, three *ab initio* calculations using different basis sets [6, 7, 8] overwhelmingly demonstrate that there is no stable axial conformer **2**. In view of this [7], the original data were reinterpreted [5] and a better fit was found for a new assignment than was found by Carreira and Lord [3].

Figure 3 shows how scattered are the equilibrium values of dihedral angle φ of azetidine reported in different experimental [9, 8, 10] and theoretical [6–8]. Up to now, theory helps here different experimental interpretations.

Bicyclo [3.1.0] hexane: The five membered ring in BCH **3** can be considered as a pseudo-four-membered ring due to the presence of a fused three-membered system. Thus, there

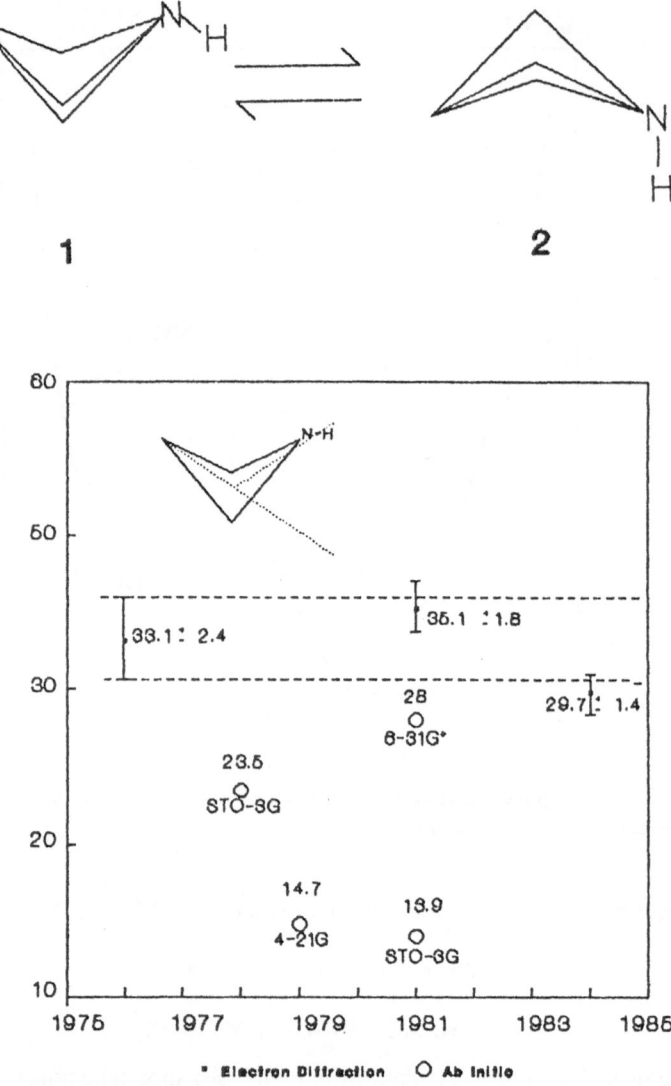

Fig. 3. Comparison of experimental [9, 8, 10] and predicted [6–8] values for the dihedral angle φ of azetidine, published in the 1976–1984 period.

is a great similarity between the four-membered ring in azetidine and the five membered fragment in BCH [1].

Lord and Malloy have interpreted far infrared spectra of BCH in terms of an SM function [4] which was confirmed later after the observation of two additional transitions in the Raman spectrum [11]. The first *ab initio* calculation [12] seemed to support this conclusion.

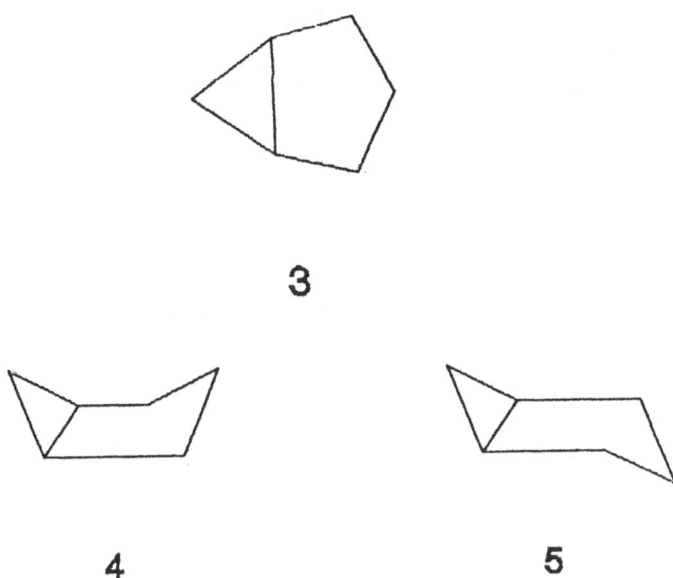

3

4   **5**

Later *ab initio* calculations [13, 14] showed, however, that the DM-function corresponded better to BCH with two possible conformers, boat **4** and chair **5**, the latter being less stable.

In fact, there is no contradiction between experiment and theory since the experimental data of Lewis *et al.* [11] extend up to ca. 1300 cm^{-1} (3.7 kcal/mole) while Okazaki *et al.* [14] report a value for the barrier to boat-chair interconversion of 3.5 kcal/mole. It means that the upper part of the potential curve of BCH deserves a special consideration.

Finally, chair conformers were found to exist in such related systems as cyclopentene oxide [4], 3,3-diethyl-6,6-diphenyl-1-azabicyclo [3.1.0] hexane [15] and *cis* 3-chloro-bicyclo [3.1.0] hexane [16].

3. Geometry of Silyl Ethers and Disiloxanes

Geometrical structures in the gas phase of molecules having the C-O-C and Si-O-Si fragments are rather well known. Less information exists, however, for the mixed, less symmetrical systems with the C-O-Si linkages. Recently, we have studied bia gas-phase electron diffraction, two molecules of this class of compounds, i.e., Me-O-SiHMe$_2$ [17] and Me-O-SiClMe$_2$ [18], and we also performed molecular mechanics calculations [19] with the previously reported force field [20] of the expected model geometry.

After this work was completed, the paper by Shambayati *et al.* [21] was brought to our attention. It's titled the *Structure and Basicity of Silyl Ethers: A Crystallographic and* ab initio *Inquiry into the Nature of Silicon-Oxygen Interactions* and it contains the results of *ab initio* calculations for a number of molecules previously studied by electron diffraction, but stops hort of making any comparisons. Thus, we found it very interesting to do a side-by-side comparison of the experimental structures with those calculated by molecular

TABLE I

Absolute and relative (in %) differences between experimental and theoretical geometries for a number of molecules studied by electron diffraction (ED), molecular mechanics (MM) and ab initio (AI) SCF.

		MeOSiHMe₂	MeOSiClMe₂	O(SiHMe₂)₂
∠	Si-O-X (deg.)	125.2(15)	127.2(21)	148.4(9)
	ED-MM[a]	+2.6	+4.5	+4.1
	Diff. (%)	2.1	3.5	2.8
	Ref.	[17]	[18]	[22]
		MeOSiH₃	MeOSiMe₃	O(SiH₃)₂
∠	Si-O-X (deg.)	120.6(10)	122.5(6)	144.1(8)
	ED-AI[b]	−4.4	−10.9	−26.0
	Diff. (%)	3.6	8.9	18.0
	Ref.	[23]	[24]	[25]

[a] Ref. [19]; [b] Ref. [21].

TABLE II

Average relative differences (%) [(exp.-theor)/exp.] for molecules containing silicon and for a hydrocarbon (C_8H_{16}).

	C-O-Si Si-O-Si		C_8H_{16}	
	MM	AI	MM	AI
Distances (%)	0.6	1.0	0.2	0.5
Bond angles (%)	2.8	10.2	1.2	1.0
Torsion angles (%)	- -	- -	4.9	4.1

mechanics and/or *ab initio* methods.

Table I shows how the comparison between experimental and theoretical values was done, by using as an example the key parameter of these systems, i.e. the bond angle at oxygen.

Table II summarizes the average relative differences for all geometrical parameters. Furthermore, since the major part of the discrepancy between experiment and theory was probably due to the presence of silicon, it was decided to include in Table II similar data for a hydrocarbon which can serve as a standard. Earlier we had studied cyclo octane by both electron diffraction and molecular mechanis [26] and later by *ab initio* methods [27].

From an inspection of data presented in Table II one can conclude the following:

— The discrepancy between experiments and theory for the hydrocarbon is better than for the silicon containing compounds by a factor 2 to 10.

— Generally, internuclear distances are better predicted than bond angles, while the bond angles are predicted better than torsion angles. The last statement refers only to one hydrocarbon, since similar data are not available for silicon containing compounds.

The poorer correspondence of torsion angles is probably related to a poorer predictive power for relative stability of different conformers. For a recent illustration of this phenomenon we refer to a paper published by Mack et al. and entitled *How Reliable are* ab initio *Calculations? Experimental and Theoretical Investigation of the Structure and Conformation of Chlorocarbonyl Isocyanate, ClC(O)NCO* [28].

4. Correlation between Ionization Potentials and Bond Angles in Heterocyclic Compounds

In 1977 Mastryukov and Osina [22] demonstrated by the use of gas phase structural data that C-CH$_2$-C fragments exhibit a linear dependence between HCH and CCC bond angles, according the following relationship:

$$\alpha_{HCH} = 126.1 - 0.175\,\alpha_{CCC} \tag{1}$$

where α_{HCH} and α_{CCC} are HCH and CCC bond angles, respectively.

Later, Mastryukov and Zefirov [23] tried to generalize this equation for those cases schematically shown below:

```
                                1
                          ┌────────→  C -CX₂- C
                          │     2
                          ├────────→  C -AX₂- C
          C-CH₂-C  ───────┤     3
                          ├────────→  C - C - C
                          │               ‖
                          │               X
                          │     4
                          └────────→  C - A - C
```

where $X = $ NH, O, S.

Consequently, the following empirical correlations were found to exist: (a) α_{XCX} vs. α_{CCC} (X = CH$_3$ and halogens); (2) α_{XAX} vs. α_{CAC} (A=Si, Ge, Sn); (3) frequencies of vibration ν (C = X) vs. α_{CCC} (X = CH$_2$, O); (4) ionization potential (IP) vs. α_{CXC}. In a recent paper [24] we are trying to model this later correlation by the theoretical procedures of quantum chemistry, at both the semiempirical and *ab initio* SCF levels by performing the proper geometry optimization.

Theoretical models can also help our understanding of the orbital nature of the target relationship, because ionization potentials are related to the highest occupied electronic states. This is a classic example how the nature of the electronic distribution of heteroatoms in saturated molecules is determined directly in a geometrical fashion.

The object molecules have saturated carbon rings with heteroatoms as N, O, and S. The following is a schematic structure:

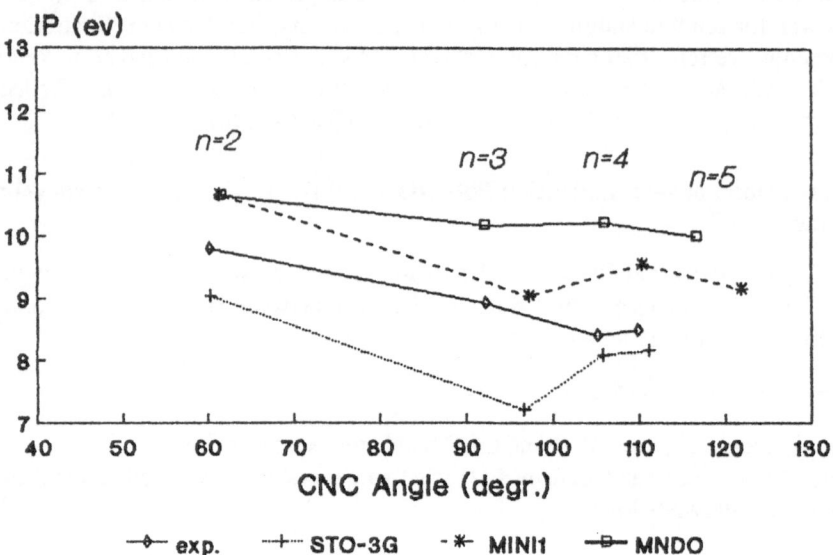

Fig. 4. Correlation of the ionization potentials (IP) and the bond angles for nitrogen containing heterocycles $NH(CH_2)_n$.

where $n = 2,3,4,5$, and $X = NH, O, S$.

Semiempirical SCF-MNDO method [25] was used, as the first approximation, to obtain minimal energy structures by the Davidon-Fletcher-Powell gradient procedure. *Ab initio* SCF calculations were performed at both minimal basis STO-3G [26] and contracted MINI1 [27] levels, with full geometry optimizations by the Murtagh-Sargent gradient method. Gradients were allowed to be lower than 0.001 (in atomic units) on all centers, except for the largest molecules ($n = 5$) where it was 0.002, with both basis sets. It has been tested that higher gradient tolerances affected both Koopman's ionization potentials and CXC angles.

It ought to be pointed out, that the SCF calculations and data processing can be done in the laboratory on IBM-PCs (or IBM comparibles) using the TC-HABANA programs for theoretical chemistry [28], which have been developed and are available for this purpose.

Figures 4 to 6 show comparisons between IP vs. bond angle relationships for all heteroatoms, in the same order.

In general, both experimental and theoretical results support the trend to "close" the angles by the sulphur heteroatom. If we take into account the third row location of this element, then the corresponding lower IP's must be caused by periodicity.

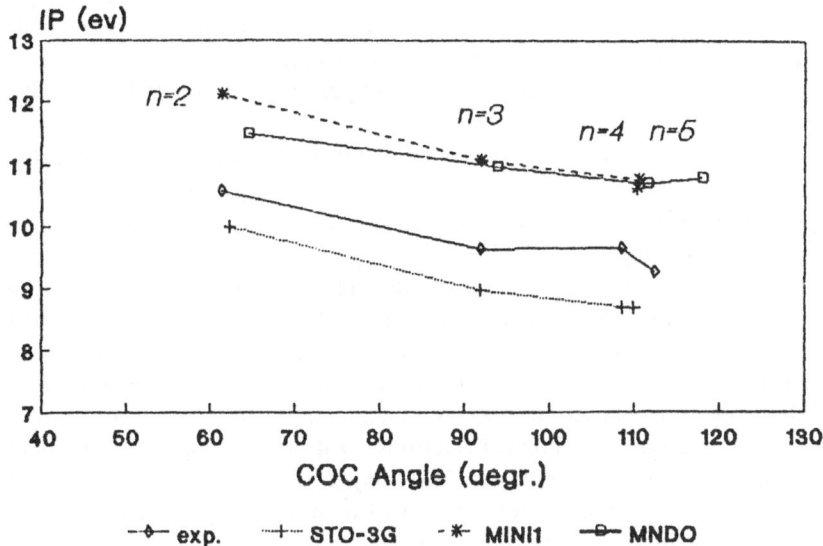

Fig. 5. Correlation of the ionization potentials (IP) and the bond angles for oxygen containing heterocycles $O(CH_2)_n$.

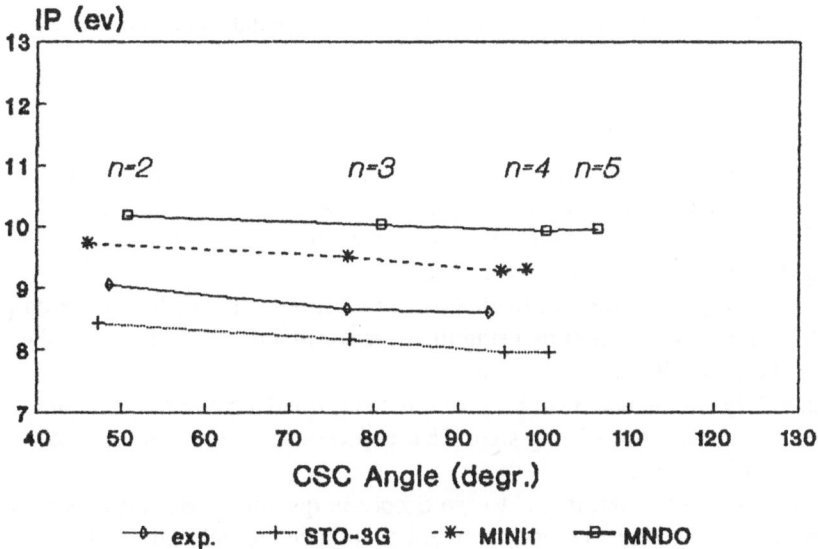

Fig. 6. Correlation of the ionization potentials (IP) and the bond angles for sulphur containing heterocycles $S(CH_2)_n$.

An interesting feature appears in the comparison between the pseudo pentagonal ($n = 4$) and pseudo hexagonal ($n = 5$) rings. In all cases the ionization potentials are very similar, with the evident exception of experimental and MINI1 values for oxygen compounds, which change significantly. Perhaps a fourth theoretical method could be assayed to investigate this anomaly, because of the importance of these particular compounds for carbohydrate chemistry.

Both ab initio results are in disagreement with the trend of nitrogen heterocycles around the $n = 3$ member of the series. This may be a case where further study is called for – either more complicated basis sets need to be used, or the accuracy of the experimental data needs to be investigated.

Even if we don't take into account the theoretical results, the relationship between experimental IP's and bond angles seems evident for all heteroatoms, confirming the previous assumptions. It is an indication about the spatial location of the highest energy state of electrons which must occur around the heteroatoms in all molecules, perhaps due to their higher electronegativity in such saturated all sigma bond rings. Organic chemists know very well that their non bonded electrons must play an important role in chemical reactions and polar associations. Particularly, oxygen heterocompounds are the basis of an important series of carbohydrates, which hydrophilic and hydrophobic regions play a determinant role in life related chemical processes.

In general, all the tested theoretical methods confirm the previous assumptions from experimental data analysis. A qualitative review of the orbital distribution by the calculated eigenvectors confirms the localization of the highest occupied molecular orbital (HOMO) mainly around the heteroatom in all molecules. Moreover, hyperconjugation of hydrogen atom electron clouds with heteroatom lone pairs occurs as in the case of cyclopropane [29], and it can explain the higher stability of their HOMO's.

Table III shows the linear regresstions in all experimental and theoretical cases, for the general equation:

$$IP = IP_O - m\, \alpha_{CXC} \tag{2}$$

where IP_O is the intercept, which would represent the energy of the electronic state around the heteroatom if a hypothetical zero degree angle could exist without a dramatical change in the orbital contribution, m is the slope, r is the absolute value of the correlation coefficient, and e_{std} is the standard error of estimation.

Paradoxically, it is easier to predict a pure electronic effect (the ionization potential from the bond angles results) by using the experimental data than by using the ab initio theoretical ones in all but oxygen cases.

Conversely, the semiempirical MNDO method, which combines quantum mechanical parametrization with statistically significant experimental results, appears to predict the expected behaviour better.

It must be pointed out, that if we go deeper in quantum mechanics, as we do when we perform ab initio calculations, more effects (which are only implicitly considered in the semiempirical methods described above) need to be taken into account. Clearly, an extended basis set would yield better results than the minimal basis STO-3G or MINI1 ones – because the variational process would have more monocentric references (as basis functions) to minimize energy. For the same reason, polarization (empty atomic state

TABLE III

Linear regressions to obtain tentative IP's from bond angles.

X	Method	IP_O (ev)	m	r	e_{std} (ev)
NH	exp.	11.64	0.0302	0.9955	0.09
	MNDO	11.27	0.0107	0.9186	0.15
	STO-3G	10.66	0.0290	0.7658	0.83
	MINI1	12.21	0.0274	0.8398	0.64
O	exp.	11.78	0.0209	0.9311	0.28
	MNDO	12.60	0.0173	0.9987	0.03
	STO-3G	11.72	0.0285	0,9835	0.18
	MINI1	13.78	0.0282	0.9855	0.17
S	exp.	9.53	0.0105	0.9630	0.09
	MNDO	10.42	0.0049	0.9998	0.00
	STO-3G	8.89	0.0097	0.9971	0.03
	MINI1	10.16	0.0090	0.9868	0.05

functions) could be included to improve results, especially in the case of the accessible empty d states of sulphur.

On the other hand, electron correlation can play a different an unpredictable role in each molecule. This effect is clearly not taken into account in our *ab initio* calculations but implicitly considered in some of the MNDO parameters which have been adjusted to fit the observed IP's and geometries.

The slopes show a general trend to decrease from NH > O > S. It appears that the ring strain is not affecting too much sulphur atom, i.e. it allows a larger flexibility for bond angles. The cause of this effect must not be attributed to the empty d levels in the heteroatom, because our *ab initio* calculations ignore them explicitly and, in spite of that, they are in fair agreement with the experimental trend.

Another facet of the overall comparison between the experimental results and theoretical models can be seen by direct comparison, as shown in Table IV. Table IV shows the average error of direct prediction for each property and the corresponding standard deviations.

In this case, the semiempirical MNDO results are clearly the worse option. The STO-3G minimal basis reproduces experimental results with few (and stable) errors, while the MINI1 basis set (whose results were better overall than the MNDO) is apparently less accurate in direct predictions, as evidenced by its high standard deviations. However, this fact is greatly influenced by the non-concordance in the bond angle of the pyperidine calculation (see Figure 4).

By way of a conclusion to this section of our work, we'd like to make the following remarks:

Our calculations reproduced the general pattern of increasing the ionization potentials with decreasing the ring size – increasing the ring strain and hyperconjunction of heteroatom lone pairs with the hydrogen atom electron clouds.

TABLE IV
Comparison between experimental and theoretical modeled results of IP's and bond angles.

X	Method	IP (ev)		Bond angles (deg.)	
		Ave. err.	Std. dev.	Ave. err.	Std. dev.
N	MNDO	1.35	0.41	2.26	3.07
	STO-3G	0.78	0.66	1.65	1.92
	MINI1	0.69	0.44	5.81	4.56
O	MNDO	1.19	0.27	3.49	1.49
	STO-3G	0.70	0.19	1.54	1.63
	MINI1	1.36	0.18	1.11	1.12
S	MNDO	1.27	0.13	5.57	3.10
	STO-3G	0.58	0.09	1.78	1.34
	MINI1	0.74	0.10	1.24	1.08
All	MNDO	1.27	0.28	3.77	2.80
	STO-3G	0.69	0.38	1.66	1.49
	MINI1	0.95	0.42	2.72	3.40

Since – in every case – the relationship between the above parameters was found to be linear, the implication is that, in principle, also other relationships, such as infrared frequencies of valence modes can be linearly correlated with geometrical features for selected families of compounds according to their orbital or electronic state dependence. Both semiempirical and *ab initio* theoretical models are fairly good at reproducing such a linearity, which confirms the electronic origin of the feature. Thus, from the viewpoint of the theoretician, Equation (2) could be written as:

$$\alpha_{CXC} = \alpha_0 + m \text{ IP} .\tag{3}$$

In that case, the correlation coefficients of Table III remain the same, but the standard errors of estimation of the bond angles become:

	exp.	MNDO	STO-3G	MINI1
N	3.08	0.84	21.89	19.46
O	12.33	1.68	6.15	5.95
S	8.65	0.74	2.64	5.67

An interesting consequence of our work for experimental structural chemists is the ease in which the theoretical calculations can be performed. Theoretical models of optimal geometries, theoretical procedures to find or confirm useful correlations, and verification of certain structural hypothesis, and other molecular geometry determinations can be performed in either the synthetic lab or at home on an IBM PC (or IBM compatible PC).

However, it must be taken into account that theoretical modesl of the geometrical structure of molecules must serve just as *models*, and not of reality. Theoretical evidence can help improve interpretations, while the corresponding methods can be improved by making comparisons with the experimental results. The reliability of any particular theoretical method to be absolutely predictive is questionable, and that includes the case of very developed *ab initio* basis function calculations, but experimental result interpretations frequently need to be supported by an independent model.

Acknowledgements

The authors are very grateful to the Universidad de La Habana and Moscow University for financial support and to UNESCO, which provided the funds for purchasing the computers used in the Havana laboratory.

References

[1] A.C. Legon: *Chem. Rev.* **80** (1980) 231.
[2] J.R. During, T.S. Little and M.J. Lee: *J. Raman Spectrosc.* **20** (1989) 757.
[3] L.A. Carreira and R.C. Lord: *J. Chem. Phys.* **51** (1969) 2735.
[4] R.C. Lord and T.B. Malloy, Jr.: *J. Mol. Spectrosc.* **46** (1973) 358.
[5] A.R. Robiette, T. Borgers and H.L. Strauss: *Mol. Phys.* **42** (1981) 1519.
[6] J. Catalán, O. Mo and M. Yañez: *J. Mol. Struct.* **43** (1978) 251.
[7] P.N. Skancke, G. Fogorasi and J.E. Boggs: *J. Mol. Struct.* **57** (1979) 259.
[8] D. Cremer, O.V. Dorofeeva and V.S. Mastryukov: *J. Mol. Struct.* **75** (1981) 225.
[9] V.S. Mastryukov, O.V. Dorofeeva, L.V. Vilkov and I. Hargittai: *J. Mol. Struct.* **34** (1976) 99.
[10] H. Günther, G. Schrem and H. Oberhammer: *J. Mol. Spectrosc.* **104** (1984) 152.
[11] J.D. Lewis, J. Laane and T.B. Malloy, Jr.: *J. Chem. Phys.* **61** (1974) 2342.
[12] P.J. Mjöberg and J. Almlöf: *Chem. Phys.* **29** (1978) 201.
[13] K. Siam, J.D. Ewbank, L. Schäfer and C. van Alsenoy: *J. Mol. Struct., THEOCHEM* **150** (1987) 121.
[14] R. Okazaki, J. Niwa and S. Kato: *Bull. Chem. Soc. Japan* **61** (1988) 1619.
[15] F.R. Ahmed and E.J. Gabe: *Acta Crystallogr.* **17** (1964) 603.
[16] M. Traetteberg, P. Bakken, R. Seip and D. Wittaker: *J. Mol. Struct.* **116** (1984) 119.
[17] O.A. Rusaeva, V.S. Mastryukov, L.V. Khristienko and Yu.A. Pentin: *Zh. Fis. Khimii*, submitted.
[18] O.A. Rusaeva, V.S. Mastryukov, L.V. Khristienko and Yu.A. Pentin: *Zh. Fis. Khimii*, submitted.
[19] O.A. Rusaeva, T.V. Timofeeva, L.V. Khristienko, Yu.A. Pentin and V.S. Mastryukov: *Zh. Fis. Khimii*, submitted.
[20] T.V. Timofeeva, V.E. Shklover and Yu.T. Struchkov: *Eksp. Teor. Khimia* **17** (1981) 674.
[21] S. Shambayati, J.F. Blake, S.G. Wierschke, W.L. Jorgensen and S.L. Schreiber: *J. Am. Chem. Soc.* **112** (1990) 697.
[22] V.S. Mastryukov and E.L. Osina: *J. Mol. Struct.* **36** (1977) 127.
[23] V.S. Mastryukov and N.S. Zefirov: to be published.
[24] L.A. Montero, S.A. Medina, A. Paneque, J.R. Alvarez and V.S. Mastryukov: *J. Molec. Struct.*, submitted.
[25] M.J.S. Dewar and W. Thiel: *J. Am. Chem. Soc.* **99** (1977) 4899.
[26] W.H. Hehre, R.F. Stewart and J.A. Pople: *J. Chem. Phys.* **51** (1969) 2658.
[27] H. Tatewaki and S. Huzinga: *J. Comp. Chem.* **1** (1980) 205.
[28] L.A. Montero, J.R. Alvarez-Idaboy, J.R. del Bosque, R. Cano, M.C. González and M.C. Rodríguez: *Folia Chim. Theor. Lat.* **16** (1988) 33.
[29] An interesting discussion of this effect from the purely experimental point of view can be read by L.V. Vilkov, V.S. Mastryukov, N.I. Sadova: Determination of the Geometrical Structure of Free Molecules, Mir Publishers, Moscow, 1983.

A THEORETICAL STUDY OF HISTAMINE MONOCATION AND SOME DERIVATIVES AS H$_2$-RECEPTOR AGONISTS

ALFONSO HERNÁNDEZ-LAGUNA and YVES G. SMEYERS
Instituto de Estructura de la Materia (CSIC)
C/ Serrano 123
28006 Madrid
Spain

ABSTRACT. A review of the main chemical, structural and biological properties of histamine is presented. Histamine and some agonists of H$_2$ receptors (4-methyl, 4-cloro and 4-nitrohistamine) are studied as monocations by means of the ab initio Hartree-Fock approximation at STO-4G level. The trans-trans structure is minimized for all of the compounds, and their geometries compared. Meaningless differences are found with substitution. Potential energy surfaces for the two more significant conformational angles are calculated, and the critical points on the potential energy hypersurface of histamine monocation are determined. A very stable structure, the so-called scorpio structure, is found in which one hydrogen of the ammonium group forms an hydrogen bond with the nitrogen of the imidazol ring. The imidazolium tautomer is found to be more stable than the histamine N$_\tau$-H tautomer. Finally, the molecular electrostatic potential (MEP) on the van der Waals surface (VDWS) is determined for histamine and the 4-substituted derivatives for some special critical points of the N$_\tau$-H tautomer as well as for scorpio structure. The range and the minimum MEP values on the VDWS of the scorpio structure is correlated with the biological activity. The molecular similarity is in agreement with the activities. Finally, an hypothetical mechanism of action is proposed for the histamine H$_2$ receptors interactions.

1. Introduction

Histamine, β-(4-imidazolyl)-ethylamine (Figure 1), is a molecule which occurs in animal as well as in plant tissues. Histamine possesses a significant biological importance because of the different actions that it produces in human organism that result in serious clinical consequences.

Histamine is an autacoide produced by enzymatic decarboxilation of L-histidine. It is either inactivated and stored, or liberated by stimuli and produce different biological actions. Histamine mediates with at least three different types of biological receptors, known as H$_1$, H$_2$ and H$_3$ histamine receptors. All of the receptors are located either on or inside the surface of the cell membrane ([1-2] and references therein). These receptors may be characterized through biological tests produced by specific agonists and antagonists. Their structure and chemical composition, however, are unknown.

Histamine produces a potent stimulation of the smooth muscle contraction. This action on an isolated ileum of guinea pig is a test which characterizes the H$_1$ receptors, as well as that of certain set agonists together with mepyramine and diphenhydramine as antagonists. Stimulation of gastric secretion and rate increasing of guinea pig atrium, as well as the action

L. A. Montero and Y. G. Smeyers (eds.), Trends in Applied Theoretical Chemistry, 149–166, 1992.
© 1992 *Kluwer Academic Publishers.*

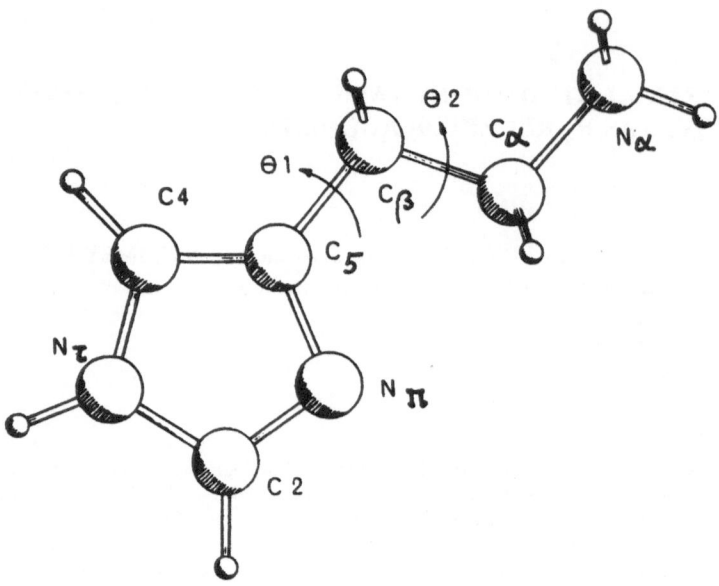

Fig. 1. N_τ-H histamine monocation. The molecule shows a σ_v plane. The atoms notation is that of Ganellin [1].

of some agonists, together with burimamide and cimetidine as specific antagonists, define the H_2 receptors of histamine. The stimulation of stomach acid secretions is associated with gastric ulcer illness. The third type of receptors of histamine, H_3, controls the release of histamine from brain slices. They are characterized by thioperamide as antagonist, and (R)-α-methylhistamine as a potent and selective agonist [3].

Some special features which could determine the biological activities of histamine could be stressed studying its chemical and molecular structure properties. Moreover, a knowledge of the molecular structure changes that influence the activity of molecules with histamine-like activity can give some insight into the active site of the receptors.

Agonists of histamine have the property of being selective with respect to different types of receptors. Hence, there exists the possiblity of extracting special structural details from the inspection of the chemical differences, to explain the effects of selectivity and activity. For these reasons, histamine and derivatives have been submitted in these last decades to extensive experimental and theoretical study in order to detect a structure-activity relationship, as well as to puzzle out the mechanisms of action regarding the different receptors ([1–2] and references therein). Theoretical studies have played an important role in the comprehension of the structure-activity relationships, in the mechanisms of action and in the form of the receptor models.

In the present paper, special attention will be paid to the H_2 action of histamine and of some of its 4-substituted agonists. We will determine conformational potential energy surfaces (PES), critical points of PES, and calculate molecular electrostatic potential (MEP) on molecular van der Waals surfaces (VDWS).

2. Chemical and Molecular Properties of Histamine. Receptor Models

In aqueous solution at low pH, histamine is found as a dication, at physiological pH (7.4), histamine appears as monocation [4]. This fact suggests that the monocation should be the chemical species which mediates with biological receptors. However, pH may be lower in the neighborhood of the cell membrane (pH = 5.4). Therefore,in its interactions with the receptors, histamine has to be considered as either a monocation or as a dication.

At physiological pH, histamine monocation exhibits three possible tautomeric forms: the N_τ-H, the N_π-H and the imidazolium tautomers. The first one presents the N_τ atom of the imidazol ring (the farthest from the ethylammonium side chain) hydrogen bonded. The second one, N_π-H tautomer, presents the N_π atom (the nearest to the side chain) hydrogen bonded. Finally, the imidazolium tautomer presents both nitrogens bonded to hydrogen atoms, with a positive charge located on the imidazolyl ring. The relative populations in water of these three forms have been estimated to be 80 %, 20 % [5] and much less than 1 % [1], respectively.

In solid state, histamine monocation is found as the N_τ-H tautomer [6], and histamine free base as N_π-H one [7]. In the monocation form, histamine is found in a configuration with an intermolecular hydrogen-bond between one of the hydrogen atoms of the ammonium terminal group and the N_π atom of the imidazol ring of the nearest molecule forming dimers. However, in the neutral form, histamine molecules appears linked as a chain through hydrogen-bonds between the nitrogen atom of the amine terminal groups and the hydrogen atom bonded to the N_τ atoms. Intermolecular hydrogen-bonds determine thus the crystalline structure of this molecule in its different chemical species.

One of the first requirements to be fulfilled by histamine-like active molecules was proposed by Walter et al. [8] and Nieman et al. [9]: the molecules should have certain special intramolecular distances between the functional groups, the possibility of chelation between the two nitrogen atoms, and being as N_τ-H-like tautomers. However, the stability of the chelated cationic form was seen to be low, at least in water solution [4]. Nevertheless, histamine forms stable complexes with divalent cations (Cu, Ni, Co, Zn and Cd) in aqueous solution by means of chelation between the N_α and N_π atoms. The stored portion of histamine in the mast cells is found likely bonded by electrostatic interactions as a Zn^{2+}-(histamine)-heparine chelate complex [10–11]. To form this chelate, histamine should be in the N_τ-H tautomer, and the N_α and N_π atoms should link two Zn^{+2} ions.

The first hypothetical distinction of the molecular structure properties with respect to the interaction with two different receptors is due to Kier [12], who determined two conformers of the histamine monocation from Extended Hückel calculations: the trans and gauche conformers. The former exhibits the ethylammonium side chain in an extended configuration (θ_2 around 180°, see Fig. 1), while the gauche conformer presents the ethylammonium side chain in a folded conformation (θ_2 around 60°). This author forwarded the hypothesis that each conformer reacts with a different receptor. Kier assigned the trans conformer to the H_1 receptors and gauche one to the H_2 receptors.

Ganellin et al. [13] determined the conformational properties of histamine in water solution, by means of NMR spectroscopy and theoretical calculations. Gauche and trans conformers are found, in 55% and 45% proportions, respectively. The conformational properties of methyl substituted histamines and their trans/gauche rates was studied against the relation of their activities H_1/H_2 [14]. 4-methyl histamine is highly selective for H_2

receptors. The crowded conformational region, hindered sterically by the methyl interaction with the side chain, should be necessary for the H_1 activity. Because of these differential conformational properties between histamine and 4-methylhistamine, a trans conformer ($\theta_1 = 0°$ and $\theta_2 = 180°$) was proposed as the essential conformation with respect to the H_1 receptors [15].

Ganellin et al. [1–2, 16] also proposed models for the interaction with the two receptors. The H_1 receptors agonists should be in monocationic form and should possess a proton on the ammonium terminal group. In the same way, the heterocyclic ring must exhibit a nitrogen atom with a lone electonic pair in ortho position with respect to the side chain, and the ring should be able either to rotate to reach dynamically a special configuration, or either to be already in an essential conformation (trans conformation with $\theta_1 = 0°$ and $\theta_2 = 180°$).

The H_2 receptors agonists should be in monocationic form with a free proton on the ammonium group. These agonists should be in N_τ-H tautomeric or in amidine form as in dimaprit [1–2, 16]. The imidazol group should be able to form a bifunctional hydrogen bond where it either it mediates in a static hydrogen bond or dynamically in a proton transfer. This last model works similarly as serine-histidine-asparte triad in certain enzymes.

Weinstein et al. [17–18] proposed a H_2 receptor model, by using ab initio calculations and molecular electrostatic potential maps (MEP), as well as the variation in the tautomeric properties of histamine monocation and the free base. The H_2 receptor structure should be a cavity with three active sites. Histamine interacts inside the cavity. In a first step, one of the active sites interacts with the ammonium group, and the other two with the N_π lone pair and the N_τ hydrogen atom. In a second step, the tautomeric preference changes from N_τ-H to the N_π-H with a proton transfer. This model is paramorphous with that of the serine protease charge relay system.

These authors determined a trans conformer with full optimization of the geometry, and indicated the existence of another gauche one. The model takes into account the existence of MEP minima on the two nitrogen atoms of the imidazolyl group. Topiol et al. [19] determined the ab initio optimal geometry of histamine and some derivatives.

Luque et al. [20] have studied the structure-activity relationship of histamine monocation and some substituted derivatives, as function of the depth of the electrostatic potential minimum on the lone pair of the N_π atom in the trans conformation. All the electrostatic potential map calculations were performed on planar cross-sections .

Most of the theoretical papers dealing with conformational calculations do not take into account the solvent effects. Pullman et al., however, have studied the conformational properties of histamine using a supermolecule hydration model into the PCILO approximation, [21]. Trans and gauche conformers were found. In contrast, no hydrogen-bond between N_π nitrogen atom and the ammonium side chain was encountered in the gauche conformation.

Kimura et al. [22] proposed an experimental H_2 receptor model. The crown amines are known to possess very strong basic properties, in such a way that they are able to trap protons inside the crown. The crown amines exhibit in some way similar properties to those of the H_2 receptors. The triprotonated macrocyclic hexamine (18)-ane-N_6 can store three protons inside the crown, The N_τ-H histamine tautomer, as monocation in aqueous solution can react with this system giving a complex. In this complex, the N_α and N_π atoms of histamine are bonded to the proton system. Histamine releases a proton from its ammonium group, consequently the pH of medium decreases.

This experimental model works in a similar way as histamine does with the H_2 receptors in the organism stimulating acid secretion. In addition, standard H_2 receptor antagonists, such as cimetidine, reacts with the complex histamine-triproton-crownamine in a competitive way, deplacing histamine and blocking the three-proton active site. The antagonists of the H_2 receptors of histamine are expected to mediate in a similar way.

Smeyers et al. [23–26], using CNDO/2 potential energy surface calculations for some H_2 receptors agonists of histamine substituted in position 4, found two minima corresponding to trans and gauche conformers. The lowest gauche conformer exhibits an internal hydrogen-bond. Moreover, these authors put forward the existence of a correlation between the mean net charges on the imidazol ring and the H_2-receptor activities in some of the trans conformers. These were precisely those in which the weakest internal bond was detected. This correlation suggested that solvent effects could open the hydrogen bond and stabilize the trans conformation in water. In such a way, an open conformation, nearly trans, was proposed as essential for the H_2 activity.

3. Methods

3.1. POTENTIAL ENERGY SURFACE DETERMINATIONS

Restricted Hartree-Fock (RHF) calculations were performed to determine the potential energy surfaces of histamine, 4-methyl, 4-chloro and 4-nitro histamine derivatives at STO-4G minimal basis set level. This basis set can be used in this series of molecules because it contains mainly hydrogen, carbon, oxygen and nitrogen atoms which possess a number of electrons in accordance with the size of the basis set. Minimal basis set indeed may be used at least for qualitative discussions, [27], whenever the theoretical results can be matched by experimental values. Extended basis sets have been used in some cases in order to check some theoretical results. On the other hand, the STO-4G basis set was preferably used instead of the STO-3G one, because that basis is expected to better reproduce the hydrogen-bonds [28].

All the calculation were performed with the MONSTERGAUSS program. [29–30]. Minimal structures were reached by some optimization methods, which were left unchanged during the conformational surface determination.

3.2. CRITICAL POINT LOCALIZATION

For the geometry optimization, the Broyden-Fletcher-Goldfar-Shano [31] and Davidon's optimally conditioned procedures were used [32]. Saddle points have been determinated by means of the VA05D procedure [33]. All of them were included as subroutines in the MONSTERGAUSS program. They use analytical gradients of the RHF energy [34–35]. The force constants, however, were calculated numerically. The stationary point orders were verified by means of the canonical curvatures obtained by diagonalization of the force constant matrices. The calculations were performed up to obtain forces in the internal coordinates of 5×10^{-3} mdyn or mdyn Å/ rad. Nevertheless, generally the mean force value was of one order of magnitude lesser.

The procedure to reach the saddle points on the potential energy surfaces resort to a division of the internal coordinate space into two subspaces: the control [36–37] and complementary subspaces. The first is formed by the three conformational angles and internal coordinates whose analytical forces change significantly with the conformational variations. So the control space is formed by the θ_1, θ_2, θ_3 and C_5-C_β-C_α and C_β-C_α-N_α.

At first stage, rigid rotor conformational potential energy surfaces were determined. The optimum geometries were obtained with a full minimization of the trans-trans conformers of histamine and derivatives. Networks were obtained by calculating the energy as a function of θ_1, θ_2 for each 30°. The potential energy maps were obtained by fitting with the help of DI 3000 package [38].

As it is easy to demonstrate from the non-rigid symmetry properties of this molecule, the trans-trans conformation is an extremum on the potential energy surface [39]. Developing the potential energy expression as a function of the rotation angles in a Fourier expansion [40], and applying the condition of extremum to the first derivative, it is found that when $\theta_1 = 0$, π, 2π, ..., and $\theta_2 = 0$, π, 2π, ... the conformation is necessarily an extremum. The trans-trans conformation is thus *a priori* a stationary point. For this reason, this conformation was the first calculated in the potential energy surface of histamine and derivatives. Furthermore, because the symmetry planes existing in the imidazolyl group and the two rotors of histamine, the conformational potential energy function of histamine is seen to be invariant under a simultaneous sign change of all the rotation angles:

$$V(\theta_1, \theta_2, \theta_3) = V(-\theta_1, -\theta_2, -\theta_3) \tag{1}$$

when the origin of the rotational coordinates are lying in the symmetry molecular plane.

In this conditions, the potential energy surface presents a two-fold symmetry axis, and only half of the network points have to be calculated [41].

3.3. MOLECULAR ELECTROSTATIC POTENTIAL ON THE VAN DER WAALS SURFACE

The molecular electrostatic potential (MEP) produced by a molecule in a certain nuclear configuration at a given point of the space is written as:

$$V_{MEP}(\mathbf{r}) = \sum_A \frac{Z_A}{||\mathbf{r} - \mathbf{R}_A||} - \int_v \frac{\rho(\mathbf{r}')}{||\mathbf{r} - \mathbf{r}'||} d\mathbf{r}'$$

where Z_A is the nuclear charge of the nucleus A, $\mathbf{R}_A$ is the corresponding position vector, $\rho(\mathbf{r}')$ is the one particle electron density function. The MEP has been calculated over the network points of the van der Waals surface (VDWS), using the GAUSSIAN 80 program [42]. The VDWS was calculated with the Connolly's MS program [43] with some modifications [44] in order to calculate variable point densities.

The VDWS's have been defined using the Gavezzoti's van der Waals radii [45]. The PEM on the VDWS has been computed on a network of points on each atomic sphere with a density of 15 points/Å^2. The MEP has been study as a function of increasing values on the whole VDWS and on the atomic spheres.

4. Results and Discussions

4.1. CONFORMATIONAL CALCULATIONS

The STO-4G geometries of histamine, 4-methyl, 4-chloro and 4-nitrohistamine monocations as N_τ-H tautomers and in the trans-trans conformation ($\theta_1 = 180°$, $\theta_2 = 180°$) are

TABLE I

Theoretical and experimental [6] bond lengths (in Å), bond angles (in degrees) of the trans-trans histamine (Hist) and 4-substituted derivatives of the N_τ-H tautomer.

	Hist		4-CH$_3$-Hist	4-Cl-Hist	4-NO$_2$-Hist
	Theor.	Exp.	Theor.	Theor.	Theor.
C2-N_π	1.319	1.328	1.317	1.321	1.327
N_τ-C2	1.379	1.349	1.380	1.380	1.375
C4-N_τ	1.386	1.362	1.389	1.387	1.386
C5-C4	1.352	1.352	1.357	1.355	1.358
N_τ-H	1.021	0.91	1.021	1.022	1.024
C_β-C5	1.521	1.498	1.520	1.519	1.519
C_α-C_β	1.541	1.509	1.540	1.541	1.542
N_α-C_α	1.535	1.505	1.536	1.535	1.533
N_τ-C2-N_π	111.8	110.0	111.7	112.1	112.2
C4-N_τ-C2	107.0	108.4	107.5	106.2	106.0
C5-C4-N_τ	105.6	106.1	104.8	106.6	107.0
H-N_τ-C2	126.6		126.7	127.4	128.6
C9-C5-C4	128.8	128.9	128.3	128.5	128.4
C_α-C_β-C5	108.1	110.8	108.0	108.1	108.5
N_α-C_α-C_β	111.5	110.3	111.6	111.4	111.0

presented in Table I. The experimental X-ray geometry values for histamine monocation are gathered in the same table [6]. All the calculated geometries have undergone a full geometry optimization in this fixed conformation. The theoretical values compared with the experimental ones are in reasonable agreement, although the ethylammonium side chain conformations are not the same [6]. Indeed the conformation in crystal are influenced by the intermolecular forces, especially when hydrogen-bonds exist, as in histamine monocation occours. The root mean square deviations (rms') between the theoretical and experimental structures are 0.045 Å and 0.6°. As to be expected, the C_α-N_α bonds lengths are found to be too large when compared with the experimental ones [46]. Our calculations were performed indeed in vacuo whithout any interactions with the surrounding medium, while histamine is forming dimers in crystal. So, the C_α-N_α bond involved in an intermolecular hydrogen bond may be expected to be shorter in the experimental structure.

Furthermore, it is well known that STO-4G geometries are in average roughly different by 0.05 Å and 10° for bond lengths and bond angles, respectively [47]. As a result, inside this approach, a reasonable agreement with the known experimental values has been reached, especially regarding the bond angles. Thus, this basis seems to be suitable to study the geometry of histamine and derivatives for our purposes: to study the relative differences in geometry due to the sustitution.

The optimized trans-trans structures have been also calculated for 4-substituted derivatives, see Table I. The geometrical differences of the theoretical structures between histamine and the derivatives are very small, with the exception of the bond angles of the nitro derivative. Anyway, these differences do not seem to be very significant. The rms' of the calculated structures of histamine and derivatives are given in Table II. The largest rms cor-

TABLE II

Root-mean-square deviations (r.m.s.) of bond lengths and angles of N_τ-H histamine and 4-substituted derivatives.

	Bond lengths (Å)	Bond angles (°)
Histamine	0.045[a]	0.6[a]
4-CH₃-Hist.	0.002	0.4
4-Cl-Hist.	0.002	0.6
4-NO₂-Hist.	0.004	1.0

[a] r.m.s with the experimental values [6].

responds to the nitro derivative, twice those of the other derivatives. No structure-activity relationship between the geometrical data and biological activity is observed.

In addition, it is seen that the deviations are very small, so the substitution on position 4 does not affect seriously the molecular geometry of the imidazol ring, as well as that of the ammonium side chain, at least at this level of approximation. Therefore, these differences do not represent a large structural variation to justify some H_2 receptor activity differences. If the essential conformation is the trans-trans, it could be inferred from these previous calculations that histamine and these derivatives could be docked on the H_2 receptor approximately in the same way, i.e., with the same probability without taking into account the possible hindrance of the substituents in position 4, that could be apreciable.

In Figure 2, the θ_1, θ_2 potential energy surface of N_τ-histamine determined with the optimized geometry given in Table I, without any additional relaxation. The trans-trans conformation appears in the center of the map. The symmetry conditions (1) are fulfilled with respect to this center. The small asymmetries observed in the map are due to the inter-polation method used to calculate the contour lines. Two conformers are found: a gauche and trans ones. The first is 11.12 kcal/mol more stable than the second. This difference of energy is very large and probably due to the minimal basis set employed. Nevertheless, it is qualitatively sigificant, especially for verifying the conformational differences in the case of the derivatives. The trans-trans conformation appears as a transition states between the two trans minima.

The conformational maps of 4-methyl and 4-chloro present a similar topography to that of histamine. The conformational areas and potential and transition structures are approximately at the same positions. The gauche conformers of the histamine, 4-methyl and 4-chloro histamines correspond to $\theta_1 = 120°$ and $\theta_2 = 60°$. These are conformers with the ethylammonium side chain folded forming a hydrogen bond between the N_π and one of the hydrogen atom of the ammonium group. On the contrary, the surface of the nitro derivative looks very different. The nitro derivative possess a new minimum due to a strong intramolecular hydrogen bond between the nitro group and the ethylammonium side chain [39].

4.2. CRITICAL POINTS ON THE POTENTIAL ENERGY HYPERSURFACE

In this previous study, the molecular forces were minimized only in the trans-trans conformation. The conformational surfaces were thus determined in the rigid rotor approximation,

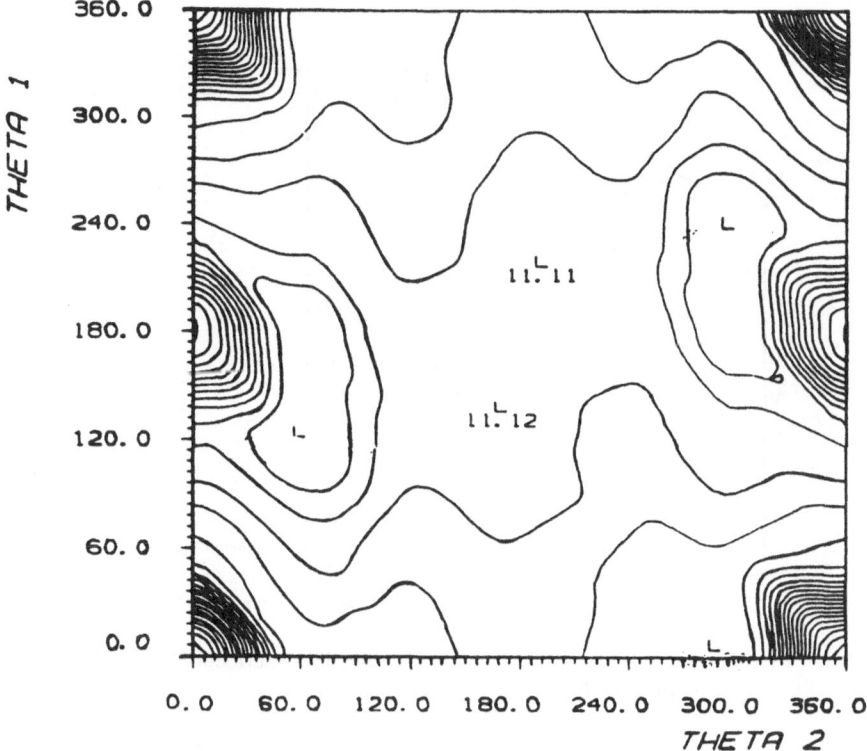

Fig. 2. Conformational map of N_τ-H histamine monocation. Contour interval 0.5 kcal/mol.

and no more critical points have been considered. Next, all the critical points situated on the potential energy hypersurface of the control space of N_τ-H histamine will be determined by minimization of all the internal coordinates.

In the trans region, we found four critical points [48]: three transition states and one minimum, (see Table III, in this table some geometrical features of the critical points are given). The minimum, M2, coincides with that found by Weintein et al. [17]. It is the so called trans conformer. The trans-trans critical point before mentioned is in fact a transition state, TS2. This transition structure connects the minimum M2 with its symmetric image obtained by inversion around the trans-trans symmetry point (the triple switch operation) [41]:

$$\hat{V}V(\theta_1, \theta_2, \theta_3) = V(2\pi - \theta_1, 2\pi - \theta_2, 2\pi - \theta_3)$$

The canonical curvature at this point is very low, (see Table IV), as to be expected because of the small difference in energy with respect to the minimum, M2, of 0.3 kcal/mol. As a result, this region of the potential energy hypersurface is seen to be very smooth.

The transition vector determined at this point in the control space is essentially a linear combination of the two torsional angles of the ethylammonium chain (see Table IV), being θ_1 the most relevant internal coordinate. Nevertheless, to overpass this critical point both conformational angles are involved in the maximization of the energy.

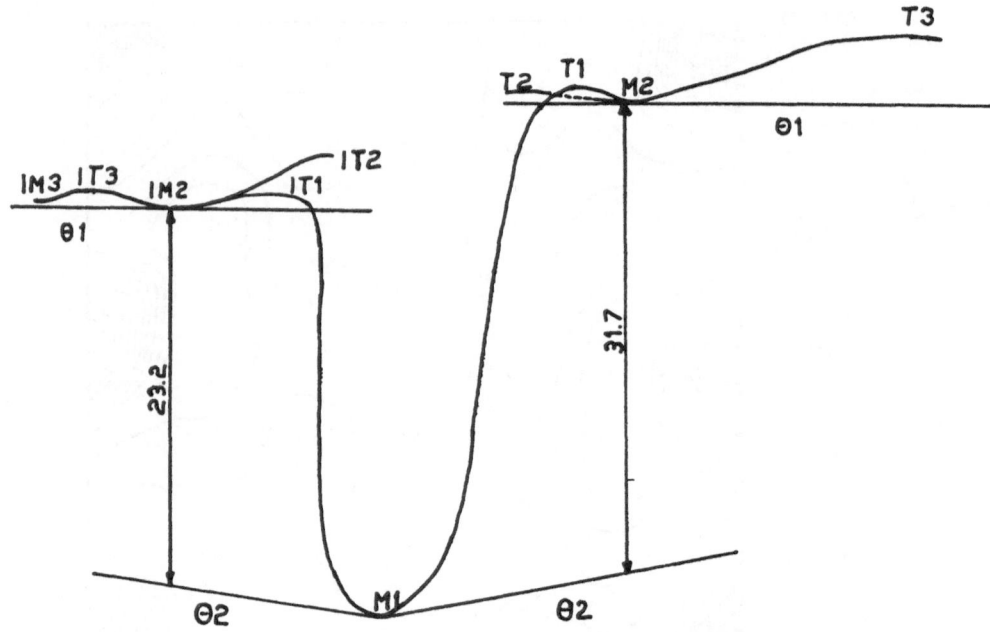

Fig. 3. Partial schematic representation of the potential energy hypersurface of N_τ-H and imidazolium tautomers of histamine monocation.

Another critical point in the trans region is the transition state, TS3, connecting the minimum M2 with its image through a rotation in the opposite sense:

$$\hat{V} V(\theta_1, \theta_2, \theta_3) = V(-\theta_1, -\theta_2, -\theta_3)$$

The energy increase with respect to this minimum is found to be 5 kcal/mol. The transition vector at this point presents a similar behavior as that of the transition structure, TS2. The mixing coefficients on θ_1 and θ_2 coordinates have now opposite signs.

Taking into account the very low curvature existing between TS2 and M2, most of the motions in the trans region should pass throughout the TS2 transition state.

The last critical point found in the trans region is the transition state, TS1, connecting this region with the gauche one. The increase in energy with respect to the M2 minimum is only of 0.8 kcal/mol. In spite of this small difference with respect to TS3 barrier, the canonical curvature is larger because of the proximity of a deep gauche conformation, $M1^c$. The transition vector depends only significantly on the θ_2 coordinate. We have to remark that θ_3 angle does not play any relevant role in the transition vectors.

Finally, a deep minimum is found for the preferred conformation, $M1^c$, starting from the TS1 transition structure, using the Optimally Conditioned method (see Figure 3 for a schematic representation of the potential energy hypersurface). This configuration is 21.1 kcal/mol more stable than secondary minimum, M2.

In the complete space, however, the $M1^c$ conformation is not found to be a critical point of the complete potential energy hypersurface, because of strong forces appearing in the imidazolyl ring as well as the hydrogen bonds of the terminal ammonium group. Relaxing

TABLE III

Some selected STO-4G bond lengths, conformational angles, potential energies and mean gradients of histamine.

	M1	TS1	M2	TS2	TS3
$H_1(N_\alpha)$-N_α	1.278	1.039	1.039	1.039	1.039
$H_2(N_\alpha)$-N_α	1.030	1.039	1.039	1.039	1.039
$H_3(N_\alpha)$-N_α	1.030	1.039	1.039	1.039	1.039
$H(N_\tau)$-N_τ	1.025	1.021	1.023	1.021	1.021
θ_1	149.1	141.4	142.5	180.0	0.0
θ_2	50.1	132.0	171.5	180.0	180.0
θ_3	−40.3	59.3	59.6	60.0	60.0
V	0.0	32.7	31.9	32.2	36.9
$G*10^3$ [a]	0.4	2.5	4.7	3.6	1.8

[a] Bond lengths in Å, bond angles in degrees, potential energies in kcal/mol and mean gradients calculated with internal gradient in mdyn or mdyn Å/rad.

the complete geometry, a new structure, M1 [48], (not a conformer) is found, in which one of the hydrogen atoms of the ammonium group is forming a very short hydrogen bond with the N_π atom of the imidazolyl ring. Furthermore, the proton appears in the hydrogen-bond at a shorter distance with respect to the N_π atom, 1.209 Å, than with respect to N_α one, 1.278 Å. The bond order of N_α-H is reduced from 0.875 in M2 to 0.4, approximately, in M1, and the N_π-H bond order increases from 0.000 to 0.473. The net charge of the transferred hydrogen increases slightly from 0.320 in M2 to 0.347 in M1. The net charges of the other two hydrogen atoms of the ammonium group decrease appreciably from 0.323 in M2 to 0.242 in M1.

This new M1 configuration is a stationary point on the complete STO-4G potential energy hypersurface of the N_τ-H of histamine. This structure, called hereafter scorpio, belongs rather to the imidazolium tautomer than to the N_τ-H one, showing a very strong hydrogen-bond. Therefore, we shall consider next the ethylamine side chain conformations in the imidazolium tautomer, and study the corresponding conformational potential energy hypersurface.

Note that if the TS1 is the only transition structure connecting the trans region with the M1 critical point, an internal coordinate related with that hydrogen tranfer could be expected in the transition vector of the TS1. Such a coordinate is not observed. Probably, another minimum (a true conformational one) and a new transition structure connecting this possible minimum with the M1 structure could exist. Both hypothetical critical points should have low curvatures in this STO-4G surface, and therefore they are located with difficultly, even more both configurations could collapse in a shoulder. The lack of these critical points could be related with the basis uncompletness.

Starting from the M1, the proton is completely transfered to the imidazol ring for forming the imidazolium tautomer. In the trans region of the potential energy hypersurface of this tautomer, five critical points are found, namely ITS1, IM2, ITS2, ITS3 and IM3 [49]. Approximately all of them have an equivalent point in the N_τ–H region, with the exception of ITS3.

TABLE IV

Transition vector and canonical curvatures (mdyn/Å) of the transition states of N−τ-H histamine.

	TS1	TS2	TS3
C_5-C_β-C_α	−0.038	−0.003	−0.001
C_β-C_α-N_α	−0.003	−0.003	−0.001
θ_1	0.025	−0.987	−0.993
θ_2	0.998	−0.156	0.112
θ_3	0.039	0.036	0.030
ζ	−0.091	−0.002	−0.012

The ITS1 critical point connects the scorpio structure with the trans region of the Imidazolium tautomer. The θ_2 angle is the main component of the transition vector. This transition structure is connected with the IM2, a minimum with approximately the same conformation of M2. The energy difference between IM2 and M1 is 23.2 kcal/mol.

IM3 is a minimum in this region and a transition structure, TS3, in the N_τ-H region. The amino group is not symmetricaly located with respect to the neighbour CH_2 group. In Newman proyection, it shows a configuration with one of the N—H bonds located between the two C—H bonds. Another region with equivalent critical points has been found. In this second region the NH_2 group appears roughly symmetrically located with respect to the CH_2 [48–49]. This last region is higher in energy. These critical points are labelled as IMi' (i =1–2) and ITSi' (i =1–3) [48–49].

In the imidazolium trans regions the N_π—H bond is not anymore submited to the strong interaction of the amino group, and it has a normal length, approximately 1.027 Å, the same that of N_τ—H bond. Most of the charge is now located on the imidazolyl ring.

In this molecule, as in many others in organic chemistry, the acid-basic properties in gas phase are drastically different from those in solution: the imidazolium monocation tautomer appears to be 8.49 kcal/mol more stable than the N_τ—H one (in this minimal basis set calculations). The reverse results are found in solution, which were commented in the point 2 of this article.

The histamine basicity has been measured by Abboud et al. [50] in gas phase by means of Fourier Transform Cyclotron Resonance Spectroscopy. They found that histamine is −28.5 ± 0.5 kcal/mol more basic than ammonia at 303 K.

In gas phase, the basicity of 4-n-propylimidazol can be estimated to be −24.8 kcal/mol more basic than ammonia; and the n-propylamine is −14.1 kcal/mol more basic than ammonia. As a result, it is clear that the basicity of the N_π is larger than the N_α. Therefore, the value measured by Abboud [50] corresponds to an imidazolium tautomer. This value should have indeed an enhancement because of the hydrogen-bond or chelate formation. If we compare the basicity of the 4-n-propylimidazol with that of histamine, a value of −3.7 kcal/mol can be estimated as chelation stabilization. This chelation gives rise to a loss of entropy which can be stimated as 10–15 cal/mol. The stabilization entalpy of the chelation can be estimated as $-3.7 + (-12.5) \times 10^{-3}$ =−7.5 kcal/mol. This value is far from that found in our calculations: 23.20 kcal/mol. This disagreement can be attributed to the use of a minimal basis set in our calculations. The M1 structure may be overstabilized with respect to the other critical points evaluated in this surface.

As to be expected, the acid-basic properties of the N_π and N_α will change in aqueous medium in such a way that the N_τ–H tautomer will be more stable than the imidazolium one.

The geometry of the scorpio structure should correspond to the real configuration in gas phase. Nonetheless, it is clear that some geometrical features have been exagerated because of the minimal basis set used. If we take into account the basisties before mentioned, the bridged hydrogen must be bonded to the N_π with a normal length, and forming a hydrogen-bond with the N_α. The N_π—H bond will be lengthed by the hydrogen-bond. This should be strong, but perhaps not enough to built the sort of chelate as that found in our calculation.

The scorpio structure, however, shows some trend of forming chelates, and is akin with that found by Kimura et al. [22], in the complex of histamine with the triprotonate macrocyclic hexamine (18)-ane-N_6. This complex was proposed as an experimental pattern of the H_2 receptor of histamine. Hence the scorpio structure may be used as a simplified model to study theoretically the model of Kimura et al. [22]. Nonetheless, this model can exhibit other geometrical features that in the scorpio structure are not included, i. e., the N_α and N_π of histamine in the complex histamine-triproton-hexamine are able to mediate with one or two protons. The model with the interaction with two protons would account for the importance of the size of the crown to form the complex with histamine [22].

4.3. MOLECULAR ELECTROSTATIC POTENTIAL ON THE VAN DER WAALS SURFACES

If the scorpio structure is related in some way with the Kimura's experimental H_2 receptors model, it should be possible to find out some structure activity relatioships between the critical points of histamine and the 4-substituted derivatives with the H_2 activity. The hypothetical mechanism related with the Kimura's model could be the following:

histamine and agonists are in solution as N_τ–H tautomer in a mixture of trans and gauche conformations; the trans (M2) could be the essential conformation [23] which would first mediates with the H_2 receptors; afterwards the proton could be ejected and the chelate complex would be formed. Hence, the transfer from the trans conformation (M2) to scorpio structure should be related with the activity. Hernández-Laguna et al. [51] have study the critical points of the PES of the 4-substituted derivatives of histamine. They found out a correlation between the activity and the difference of energy between M2 and M1 minima. However, the methyl derivative leaves a little from the main trend and appears to be more active than histamine. For this reason, an analysis of the MEP on the VDWS has been applied to the histamine and the 4-substituted derivatives of histamine mentioned in 4.1 in the M2, TS1 and M1 critical points [52]. The other critical points were not considered in our analysis, since they are lying outside the path of this hypothetical mechanism of action.

The minimum and maximum MEP values on VDWS of each molecule are given in Table V [52]. Taking into account that all the molecules are cation, the MEP values are positive. The range of MEP is shorter on the M1 than M2 and TS1 critical points. This fact is due to the folded scorpio structure, in which the positive charge is indeed best shared throughout the surface of the molecule. The substituent has different effects on the MEP on the VDWS depending of the critical point: in M1 the minimum value drops down along the family and the maximun increases, with the exception of the methyl derivative; TS1 and

TABLE V

Ranges of MEP (in atomic units) on the VDWS and pharmacological activity (% relative to histamine) for histamine and its 4-substituted derivatives on M1,T1 and M2 critical points.

	MEP min.	MEP max.	Activity [a]
Hist (M1)	0.1355	0.2132	100
4-CH$_3$-Hist (M1)	0.1079	0.2088	43
4-Cl-Hist (M1)	0.0807	0.2271	11
4-NO$_2$-Hist (M1)	0.0355	0.2360	0.6
Hist (T1)	0.0431	0.2820	
4-CH$_3$-Hist (T1)	0.0421	0.2788	
4-Cl-Hist (T1)	0.0534	0.2851	
4-NO$_2$-Hist (T1)	0.0212	0.2886	
Hist (M2)	0.0305	0.2848	
4-CH$_3$-Hist (M2)	0.0350	0.2823	
4-Cl-Hist (M2)	0.0498	0.2862	
4-NO$_2$-Hist (M2)	0.0205	0.2886	

[a] The activities were determined 'in vitro' on Guinea-pig atrium in presence of propanolol [1–2].

M2 have a wider MEP range than M1 and the maximum values of the extrema are on the Chloro or Nitro derivatives. Note that the minimum value and the range of MEP increase uniformly along the family of the compounds in M1. The uniform changes fail in the other critical points, M2 and TS1.

If we compare the range of the MEP with the biological activity, the broader is the range of the MEP the smaller is the activity, and the lower is the minimum value of MEP the smaller is the activity. The minimum value of MEP on the VDWS of the molecules in the scorpio structure shows a significant correlation with the natural logarithm [53] of the biological activities of the histamine and these derivatives, Fig. 4, the correlation coefficient is 0.989. The maximum presents a correlation too, but less meaningful: the correlation coefficient is 0.921. The other critical points do not show a suitable correlation whenever the extremal values of MEP are considered.

The localization of the minimum values of MEP on the VDWS in the M1 occurs in the neighbourhood of the atomic sphere of C$_4$ where the substituent lies. For histamine, the minimum is located on C$_4$ and on the lateral chain carbon atoms (C$_9$ and C$_{10}$). For the derivatives, the minimum is located on the C$_4$ and on the atomic spheres of the substituent. These positions are far from the hypothetical ones involved in the interaction with the H$_2$ receptors: those of the atoms in the hydrogen bond region.

From the study of the topological space created by the VDWS and the MEP, using the Euler-Poincaré characteristic for the differents ranges of MEP's, G. Arteca et al. [52] showed that the degree of molecular similarity is in agreement with the activities:

Hist == CH$_3$-Hist = Cl-Hist – NO$_2$-Hist

where ==, = and – indicated the degree of decreasing molecular similarity for the Euler-Poincaré vectors. This was found in the range of MEP of 0.143–0.160 AU for the M1 configurations. This MEP range is located on the more characteristic atoms, especially on

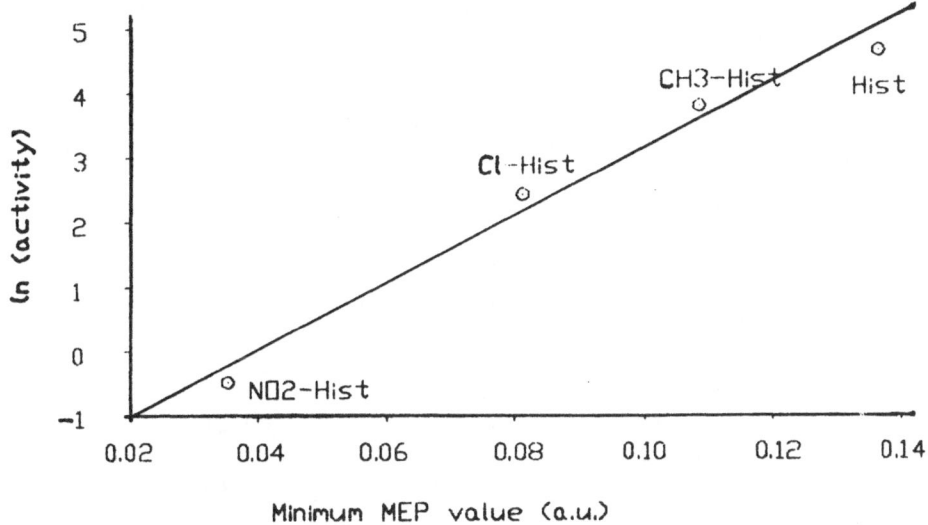

Fig. 4. Natural logarithm of the activity (the activities were determined 'in vitro' on Guinea-pig atrium in presence of propanolol [1–2]) as a function of the minimum MEP values on the VDWS of M1 structures for histamine and 4-substituted derivatives.

N_π.

N_π should be also an important atom in the formation of the complex. In Table VI, the ranges of the MEP on the N_π spheres for the three critical points are shown. Once again, as a general trend, for the M1 structure, the larger the potential on the N_π atom the smaller the activity of the compound. If we take into account this effect, once again, the methyl derivative seems deviate from this rule and appears to be more active than histamine.

From Table VI, the following remarks are possible: a) the range of MEP on the particular atomic sphere is highly dependent on the configuration; b) the minimum values on the TS1 and M2 are nearly coincidents with the overall minimum of MEP (with the exception of the nitro derivative); c) the electrostatic potential on the N_π atom increases for the M2 to M1 transformation; and d) the minimum on M2 configuration increases in the series (with the exception of the nitro derivative).

From the methyl derivative electronic properties a larger biological activity than that of histamine should be expected. Nevertheless our study is very simplified and only one hydrogen bond is considered, and no environment effect of the receptor and solvent are taken into account. If this model were correct, the role of the methyl group would be realized taking those effects in consideration.

From all the results above discussed, an hypothetical model could proposed. The mechanism of action of agonists of histamine would be a process in three steps: a) the substrat, as N_τ-H tautomer, and as cation in an open conformation, approachs the receptor, and recognizes the active site [52]; b) histamine mediates with the cationic site in a certain protein cavity, as a result, the acid-base properties of N_π and N_α change and the H^+ is ejected; and c) histamine complexates and adopts a chelated structure [52]. In the first step, the recognition by the cationic site of the receptor should be stronger in the position

TABLE VI

Ranges of MEP (in atomic units) on the N_π for histamine and its 4-substituted derivatives on M1,T1 and M2 critical points.

	MEP min.	MEP max.
Hist (M1)	0.1470	0.1990
4-CH$_3$-Hist (M1)	0.1424	0.1936
4-Cl-Hist (M1)	0.1532	0.2103
4-NO$_2$-Hist (M1)	0.1593	0.2189
Hist (T1)	0.0432	0.1295
4-CH$_3$-Hist (T1)	0.0417	0.1278
4-Cl-Hist (T1)	0.0534	0.1414
4-NO$_2$-Hist (T1)	0.0599	0.1537
Hist (M2)	0.0306	0.1183
4-CH$_3$-Hist (M2)	0.0350	0.1181
4-Cl-Hist (M2)	0.0498	0.1341
4-NO$_2$-Hist (M2)	0.0486	0.2886

of lower MEP (which is found in open conformation on the N_π atomic sphere). The last step would be the determining feature of the subsequent biological activity, since the correlation was found only for the scorpio structures. In all the cases, along the three steps, the features establishing the degree of molecular similarity can be determined by the shape characteristics found in the range of MEP on the N_π atom.

5. Conclusions

From these studies we have put forward the role of the substituent in the molecular geometry of the trans-trans histamine monocation as a N_τ-H tautomer in gas phase. We have found that the substituent does not introduce an important geometric distortion in the molecular geometry. Hence, all the studied derivatives have approximately the same chance to be docked on the active site of the receptor from the geometrical point of view. Of course, the substituent may have some hindrance effect, impossible to be detected from these calculations.

The stationary points of the PES of histamine have shown this molecule to be a imidazolium tautomer in gas phase with an internal hydrogen bond. The well known tautomers in solution are definitively very different from the gas phase tautomers. As a result, the conformation and tautomers found in water solution are modified because of the interactions with the solvent molecules. However, a question does arise: If the substrate is located into the active site of the H_2 receptor in a nonaqueous cavity, will histamine have the same tautomeric properties?

The structure found at STO-4G level is presented as a chelate, being the proton the coordination center, the so-called scorpio structure. This structure is akin to that found by Kimura in their experimental H_2 receptor model. The structure of the Kimura's model should be more complicated than the theoretical structure, which possesses only a coordination center. Moreover, this experimental receptor will be much simpler than the biological one.

So, the scorpio structure is an oversimplification of the real structure in the active site, if this receptor is the right one.

Taking into account, furthermore, the correlations of MEP and the molecular similarities of the 4-substituted derivatives, a hypothetical mechanism of action in three steps might be proposed: a) the agonist approachs the active site and recognizes it; b) the interaction with the active site lets the tautomeric properties to be changed and ejects the proton; and c) the agonist is complexated with the cationic site and the biological action is triggered.

References

[1] C.R. Ganellin and G.J. Durant: *Burger's Medicinal Chemistry*, III, M.E. Wolff (Ed.), John Wiley and Sons, New York (1981).
[2] C.R. Ganellin: *Pharmacology of Histamine Receptors*, C.R. Ganellin and M.E. Parson (Eds.), Wright. PSG, Bristol (1982).
[3] J. M. Arrang, M. Garbarg, J.-M. Laconte, H. Pollard, M. Robbe, W. Schunack, and J. C. Schwartz: *Nature* **327**, 117 (1987).
[4] T. B. Pavia, M. Tominaga, and A. C. M. Pavia: *J. Med. Chem.* **13**, 689 (1970).
[5] C. R. Ganellin: *J. Pharm. Pharmacol.* **25**, 787 (1973).
[6] K. Prout, S.R. Critchley, and C.R. Ganellin: *Aca Cryst.* **B30**, 2884 (1974).
[7] J.J. Bonet and J.A. Ibers: *J. Am. Chem. Soc.* **95**, 4829 (1973).
[8] L. A. Walter, W. H. Hunt, and R. J. Fosbinder: *J. Am. Chem. Soc.* **63**, 2771 (1941).
[9] C. Niemann and J.T. Hays: *J. Am. Chem. Soc.* **64**, 2288 (1942).
[10] L. Kerp: *Int. Arch. Allergy Appl. Immunol.* **22**, 112 (1963).
[11] M. F. Murphy, V. B. Haarstad, and F. B. Hahn: *Int. J. Quantum Chem.: Quantum Biology Symp.* **1**, 149 (1974).
[12] L.B. Kier: *J. Med. Chem.* **11**, 441 (1968).
[13] C. R. Ganellin, E. S. Pepper, G. N. J. Port, and W. G. Richards: *J. Med. Chem.* **16**, 610 (1973).
[14] C. R. Ganellin, G. N. J. Port, and W. G. Richards: *J. Med. Chem.* **16**, 616 (1973).
[15] C. R. Ganellin: *J. Med. Chem.* **16**, 620 (1973).
[16] G. J. Durant, C. R. Ganellin, and M. E. Parson: *J. Med. Chem.* **18**, 905 (1975).
[17] H. Weinstein, D. Chou, C.L. Johnson, S. Kang, and J.P. Green: *Mol. Pharmacol.* **12**, 738 (1976).
[18] S. Topiol, H. Weinstein, and R. Osman: *J. Med. Chem.* **27**, 1531 (1984).
[19] S. Topiol: *J. Compt. Chem.* **8**, 142 (1987).
[20] F. J. Luque, F. Sanz, F. Illas, R. Poulana, and Y. G. Smeyers: *Eur. J. Med. Chem.* **23**, 7 (1988).
[21] B. Pullman and J. Port: *Mol. Pharmacol.* **10**, 360 (1974).
[22] E. Kimura, T. Koike, and M. Kodama: *Chem. Phar. Bull.* **32**, 3569 (1984).
[23] Y.G. Smeyers, F.J. Romero-Sánchez, and A. Hernández-Laguna: *J. Mol. Struct. (Theochem.)* **123**, 431 (1985).
[24] A. Hernández-Laguna, F.J. Romero-Sánchez, and Y.G. Smeyers: *An. Quim.* **81**, 247 (1985).
[25] Y.G. Smeyers, F.J. Romero-Sánchez, and A. Hernández-Laguna: 'QSAR and Strategies in the Design of Bioactive Compounds', J.K. Seydel (Ed.) Verlag Chemie (1986).
[26] Y.G. Smeyers, A. Hernández-Laguna, C. Muñoz-Caro, and F.J. Romero-Sánchez: *J. Chim. Phys.* **84**, 633 (1987).
[27] S. Wilson: 'Ab Initio Methods in Quantum Chemistry', in *Advances in Chemical Physics*, LXVII, K.P. Lawley (Ed.), John Wiley and Sons (1987).
[28] J. del Bene and J.A. Pople: *J. Chem. Phys.* **52**, 4858 (1970).
[29] M.R. Peterson and R.A. Poirier: *Program MONSTERGAUSS*, University of Toronto, Ontario, Canada 1980. In addition to the GAUSSIAN 70 integral and SCF routines, the program incorporates analytic gradients and automatic optimizations with and without constrains.
[30] W.H. Hehre, W.A. Lathan, R. Dichfield, M.D. Newton, and J.A. Pople: *GAUSSIAN 70 Program*, Q.C.P.E. No. 236.
[31] C.G. Broyden: *J. Inst. Math. Appl.* **6**, 76 (1970);
R. Fletcher: *Comput. J.* **13**, 317 (1970);
D. Goldfarb: *Math. Comput.* **24**, 23 (1970);
D.F. Shanno: *Math. Comput.* **24**, 647 (1970).

[32] W.C. Davidon and L. Nazareth: 'OC program; Algorithm', is describe by W.C. Davidon, *Mathematical Programming* **9**, 1 (1975).
[33] M.J.D. Powell: *VA05 Program*, Harwell Subroutine Library, Atomic Energy Research Establishment, Harwell, United Kingdom.
[34] P. Pulay, G. Fogarasi, F. Pang, and J.E. Boggs: *J. Am. Chem. Soc.* **101**, 2550 (1979).
[35] H.B. Schlegel: *Program Force*, PhD Thesis. Queen's University, Kingston, Ontario, Canada (1975).
[36] J. Andres, R. Cardenas, E. Silla, and O. Tapia: *J. Am. Chem. Soc.* **110**, 666 (1988).
[37] O. Tapia and J. Andres: *Chem. Phys. Lett.* **109**, 471 (1983).
[38] DI3000 (c): Precision Visuals Inc. (1984).
[39] Y. G. Smeyers, A. Hernández-Laguna, J. J. Rández, and F. J. Rández: *J. Mol. Struct (Theochem)* **207**, 157 (1990).
[40] J. Maruani, Y.G. Smeyers, and A. Hernández-Laguna: *J. Chem. Phys.* **76**, 3123 (1982); Erratum *J. Chem. Phys.* **81**, 1519 (1984).
[41] Y. G. Smeyers: *J. Mol. Struct. (Theochem)* **107**, 3 (1984).
[42] J. S. Binkley, R. S. Whiteside, R. Krishman, R. Seeger, D. J. Defrees, H. B. Schelegel, S. Topiol, L. R. Khan, and J. A. Pople: *QCPE Bull.* **13**, (1981);
U. C. Singh and P. A. Kollman: *QCPE Bull.* **117**, (1982), Prog. No. 446.
[43] M. L. Connolly: *QCPE Bull.* **75** (1981), Prog. No. 429.
[44] G. A. Arteca: *Program VDWMEP*, University of Saskatchewan, Saskatoon, Canada, 1989.
[45] A. Gavezzoti: *J. Am. Chem. Soc.* **105**, 5220 (1983).
[46] J. A. Pople: 'Application of Electronic Structure Theory', in *Modern Theoretical Chemistry*, H. F. Schaefer III (Ed.), Plenum Press, New York (1977).
[47] J. Sauner: *Chem. Rev.* **89**, 199 (1989).
[48] O. Tapia, R. Cárdenas, Y. G. Smeyers, A. Hernández-Laguna, J. J. Rández, and F. J. Rández: *Int. J. Quantum Chem.* **38**, 727 (1990).
[49] A. Hernández-Laguna, G. A. Arteca, Y. G. Smeyers, and P. G. Mezey: *Reaction Paths on the Potential Energy Hypersurface of Histamine Monocation. A Molecular Shape Analysis Approach.* to be published.
[50] J.-L. Abboud, H. Homan, and M. T. Cañada: unpublished results.
[51] A. Hernández-Laguna, Y. G. Smeyers, and J. J. Rández: unpublished results.
[52] G. A. Arteca, A. Hernández-Laguna, J. J. Rández, Y. G. Smeyers, and P. G. Mezey: *J. Comp. Chem.* **12**, 705 (1991).
[53] F. Peradejordi: 'Quimica Teorica', Vol. II, (S. Fraga), in *Nuevas Tendencias* C. S. I. C. (1989).

DELOCALIZATION MECHANISM OF FERROMAGNETIC EXCHANGE INTERACTIONS IN THE COMPLEXES OF Cu(II) WITH NITROXYL RADICALS

R.N. MUSIN, P.V. SCHASTNEV and S.A. MALINOVSKAYA
Institute of Chemical Kinetics and Combustion
Siberian Branch of the USSR Academy of Sciences
Novosihirsk 630090
Russia

In the last years there has been a large number of papers devoted to the problem of designing a novel class of ferro- (or ferri-) magnetic materials based on the coordination compounds of transition metals with paramagnetic organic ligands [1–3]. Highly promising in this respect are the recently synthesized complexes of bis-chelates Cu(II) with nitroxyl radicals (L) in which the central ion of Cu(II) is directly coordinated with the $O-N<$ group of the radical [2–9]. According to magnetostructural studies the character of exchange interactions between the paramagnetic centres $\dot{C}u(II)$ and $O-N<$ in such complexes depends on the type of Cu(II)...L coordination. For example in the equatorially coordinated complexes, the strong antiferromagnetic interactions $|J| > 100$ cm^{-1} are usually observed, whereas in the axially coordinated ones only the ferromagnetic interactions 10 cm^{-1} $\leq J \leq$ 65 cm^{-1} have been detected. Here J is the exchange parameter involved in the spin-Hamiltonian $\hat{H}_s = -J\hat{S}_{\dot{C}u}\hat{S}_L$. At present there is no clear theoretical interpretation in the literature for the magnetic (particularly ferromagnetic) properties of such compounds. Most authors assume that the main mechanism responsible for the formation of intramolecular ferro- and antiferromagnetism in the Cu(II) complexes under study is the direct interaction determined by the overlap between the orbitals of unpaired electrons (magnetic orbitals) of paramagnetic centres $\dot{C}u(3d_{xy}AO)$ and $O-N<$ (π^*-antibonding MO). However, no convincing evidence is available for the validity of this approach. All explanations are of a qualitative character and are reduced to the analysis of the peculiarities of the overlap between magnetic orbitals [3, 4].

We have carried out a detailed quantum-chemical analysis of the possible mechanisms of exchange interaction in molecular (magnetic) fragments $\dot{C}u(II)...O-N<$ (or $>N-O...\dot{C}u(II)...O-N<$) of bis-chelating complexes of Cu(II) with nitroxyl radicals and the determination of the role of various factors (electronic and geometrical) in the realization of ferromagnetic interactions. The necessary wave functions have been obtained using ab initio (program GAUSSIAN-80) [10] and INDO (program SPIN-HAMILTONIAN) [11] Hartree-Fock calculations for model complexes [Cu(II)]...L given in Table I. Figure 1

L. A. Montero and Y. G. Smeyers (eds.), Trends in Applied Theoretical Chemistry, 167–173, 1992.

TABLE I

Models of complexes [Cu(II)]...L. The method of calculation of the wave functions (MO) and the structural formulas of paramagnetic fragments: [Cu(II)] – chelating moiety of the complex, L – nitroxyl radical, R – methyl or phenyl.

Method	[Cu(II)]	L	Model
ab initio HF SCF			A
			B
INDO			C
			D

shows the view of a molecular structure of the typical axially coordinated complex (of the type Cu(hfac)$_2$TEMPOL). In our calculations we have used the experimental values of the geometrical parameters R, ϑ, ψ, φ, α, and β of the complexes [3]. The main results of our study are the following:

(i) The nonempirical calculations of the parameters $J = J_{\text{dir}}$ of the direct exchange interaction in the model complex [Cu(II)]...L performed in terms of the pair orbital approximation [12] using the ab initio RHF and UHF MO of isolated paramagnetic fragments [Cu(II)] and L (model A) testify that (due to the smallness of the contributions to J_{dir} of the terms determined by the effects of spin polarization of inner shell electrons) in order to obtain the reasonable estimates of the J_{dir} values it is quite sufficient to consider the exchange interaction of unpaired electrons only. These are mainly localized at $3d_{xy}$ AO of Cu(II) and π^* MO of the O–N< group of radical L:

$$J_{\text{dir}} \simeq 2 \cdot \rho_{xy} \cdot J(3d_{xy}, \pi^*), \tag{1}$$

$$J(3d_{xy}, \pi^*) = \langle 3d_{xy}\pi^* \mid \pi^* 3d_{xy} \rangle - 2s \cdot \langle \pi^* \mid \hat{T} \mid 3d_{xy} \rangle + \\ + s^2 \cdot [\varepsilon_{xy} + \varepsilon_{\pi^*} + \langle 3d_{xy} \mid \pi^*\pi^* \rangle] \tag{2}$$

where $\rho_{xy} \simeq 0.8 \div 0.9$ is the density of unpaired electron on $3d_{xy}$ AO; $S = \langle 3d_{xy} \mid \pi^* \rangle$ is the overlap integral between magnetic orbitals; ε_{xy} and ε_{π^*} are the orbital energies;

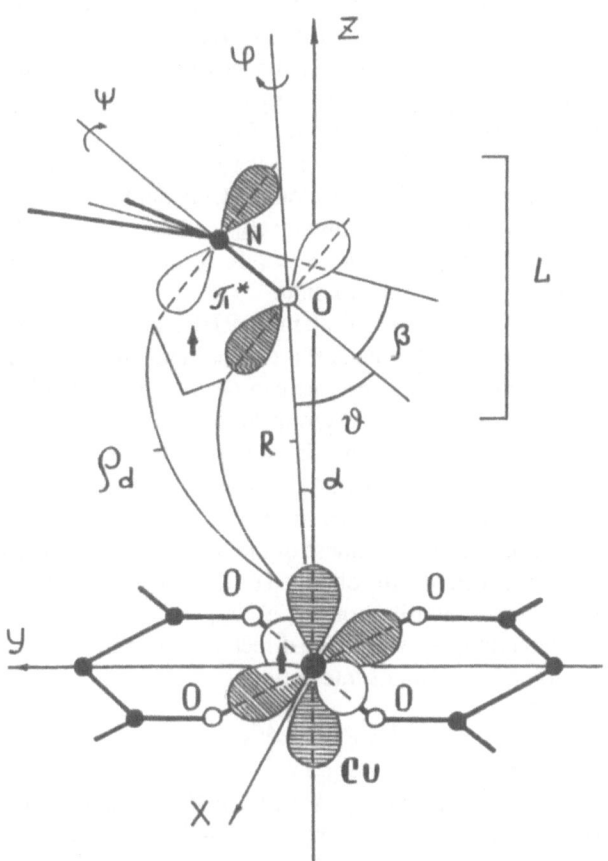

Fig. 1. The general view of the molecular structure and the main geometrical parameters of axially coordinated complexes of Cu(II) with nitroxyl radicals (L): R – the distance Cu−O(L), ϑ – the angle between the N−O(L) bond and direction Cu−O(L), ψ – the dihedral angle between the Cu−O(L)−N plane and the middle plane of radical L, angle φ determines the radical rotation around the Cu-O(L) bond (being equal to the angle between the projection of the N−O(L) bond on the XY plane and the Cu−O (chelating ring) bond), angle α characterizes the deviation of the Cu−O(L) bond from axis Z, β – the angle between the N−O(L) bond and the C−N−C (or H−N−H) bond of radical L.

$\langle 3d_{xy}\, \pi^* \,|\, \pi^*\, 3d_{xy} \rangle$ and $\langle 3d_{xy}\, 3d_{xy} \,|\, \pi^* \pi^* \rangle$ are the two-electron integrals of the exchange and Coulomb types; $\hat{T}$ is the kinetic energy operator. According to our calculations, a strong antiferromagnetism of equatorially coordinated complexes of Cu(II) can be explained by direct interaction of unpaired electrons. Indeed, these complexes are characterized by fairly large values for the overlap integral $S \simeq 10^{-1}$. In this case the last (negative) term, proportional to S^2, predominates in expression (2). This gives a large negative value for J_{dir}. The calculated values for J_{dir} are in satisfactory agreement with the experimental ones for J_{exp}. For instance, for the complex Cu(hfac)$_2$TEMPO [5]: $S = 0.07$, $J_{\mathrm{dir}} = 1.7 \times 10^3$

cm^{-1}, $J_{exp} < -450$ cm^{-1}. Different is the case for the axially coordinated complexes of Cu(II) [3, 4, 6–9]. Table II summarizes the values of S, J_{dir}, and J_{exp} for a series of such complexes. There is a substantial discrepancy with the experimental data. Despite the coincidence between the signs of exchange parameters, the calculated values for J_{dir} are about an order of magnitude less than those for J_{exp}. This is due to the fact that in such complexes the values $S \simeq 10^{-3} \div 10^{-4}$ are very small (at $\alpha = 0°$ the contribution to the overlap integral S from the orbitals of atom O(L) is negligible). Therefore in expression (2), the first (positive) term is predominant: $J_{dir} \simeq 2\rho_{xy} \cdot \langle 3d_{xy} \pi^* | \pi^* 3d_{xy} \rangle$. Thus, the observed ferromagnetic properties of the axially coordinated complexes of Cu(II) cannot be explained by direct exchange interaction.

(ii) The performed by us INDO and ab initio RHF analysis of spin density distribution in model complexes [Cu(II)]...L (Table I) shows that in magnetic fragments Cu(II)—O$\overset{\cdot}{-}$N< of the axially coordinated complexes, one can observe a slight delocalization of unpaired electron from π^* MO of the O−N< group to the valence AO of the central Cu(II) ion (Figure 1). The account of such a delocalization formally corresponds to the substitution of expression (1) by $J = 2 \cdot \rho_{xy} J(3d_{xy}, \tilde{\pi}^*)$ where $\tilde{\pi}^* = \pi^* + \sum_k c_k \varphi_k$ is a new magnetic orbital containing the admixtures of $\varphi_k = 3d, 4s$ and $4p$ AO of Cu(II) (except for $3d_{xy}$ and $3d_{x^2-y^2}$ AO because of the smallness of the overlap between these AO and π^* MO). It can readily be demonstrated that the obtained (after substitution of $\tilde{\pi}^*$) exchange parameter J will contain two main contributions $J \simeq J_{dir} + J_{del}$ corresponding to two different mechanisms of interaction, i.e., the above direct exchange interaction (J_{dir}) is followed by a new delocalization mechanism of interaction:

$$J_{del} \simeq 2 \cdot \rho_{xy} \cdot \sum_k \rho_k \cdot \langle 3d_{xy} \varphi_k | \varphi_k 3d_{xy} \rangle, \tag{3}$$

caused by the appearance on the φ_k AO of Cu(II) ion of spin density ($\rho_k = c_k$) and subsequent intratomic (ferromagnetic) exchange interaction with unpaired electron mainly localized on the $3d_{xy}$ AO of Cu(II). From our calculations it was established that the delocalization on the $3d_{z^2}$ AO of Cu(II) plays a significant role in the realization of this mechanism and moreover the smallness of delocalization magnitude ($\rho_{z^2} \simeq 10^{-2} \div 10^{-4}$) is quite compensated by a strong intratomic interaction ($\langle 3d_{xy} 3d_{z^2} | 3d_{z^2} 3d_{xy} \rangle \simeq 10^4$ cm^{-1}). Neglecting a very small contribution of exchange integrals including $\varphi_k = 4s$ and $4p$ AO and taking into account the closeness of the values of these integrals for $\varphi_k = 3d_{z^2}$, $3d_{xz}$, and $3d_{yz}$ AO, the parameter J_{del} can be represented in a more simple form:

$$J_{del} \simeq k \cdot \rho_d, \tag{4}$$

where $k = 2 \cdot \rho_{xy} \cdot \langle 3d_{xy} 3d_{z^2} | 3d_{z^2} 3d_{xy} \rangle \simeq$ const. (since in the series of the studied complexes of Cu(II), $\rho_{xy} \simeq$ const.), $\rho_d = \rho_{z^2} + \rho_{xz} + \rho_{yz}$ is the total delocalization spin density of the $3d$ AO of Cu(II).

Table II gives the values for ρ_d and J_{del} calculated from formula (3) using INDO and ab initio (in brackets) MO for the axially coordinated complexes I–VI. The values obtained for J_{del} are much higher than those for J_{dir} and, what is especially important, are in good agreement with the experimental data, namely, not only the sign ($J_{del} > 0$) but also the order of magnitude of the parameters J_{exp} are reproduced. The best agreement with experiment was achieved for the complexes I and VI in whose crystals the structurally equivalent magnetic fragments of the type Cu(II)...O$\overset{\cdot}{-}$N< (Figure 1) are realized. In the

TABLE II

Calculated values of the parameters of direct J_{dir} and delocalization J_{del} mechanisms of exchange interaction in the axially coordinated complexes of Cu(II) with nitroxyl radicals. J_{exp} – parameters obtained by magnetic measurements.

N	Compound	Direct mechanism		Delocalization mechanims		Experiment	Ref.
		$S \times 10^4$	J_{dir} (cm^{-1})	$\rho_d \times 10^2$	J_{del} (cm^{-1})	J_{exp} (cm^{-1})	
I	Cu(hfac)₂TEMPOL	0.989	2.153	0.067 (0.0.37)	11.149 (7.554)	13±5	[6]
II	Cu(pacTEMPOL)₂	1.018 0.851	1.236 0.432	0.072 0.015	12.466 2.715	19±5	[7]
III	Cu(proxFORMIL)₂	5.539	4.462	0.242	41.744	21±5	[8]
IV	Cu(hfac)₂NITMe	3.159 12.608	2.735 4.171	0.127 0.268	21.508 46.353	26±5	[9]
V	Cu(hfac)₂(NITPh)₂	7.451	4.189	0.222	38.607	10±5	[4]
VI	Cu(tfac)₂NITMe	17.324	6.847	0.468 (0.510)	81.425 (98.142)	65±5	[3]

case of complexes II–V the comparison with experiment is more difficult because the Cu(II) ions form the coordination bonds with two nitroxyl ligands L. Moreover, the crystals of compounds II and IV contain two types of magnetic fragments Cu(II)...O–N< with different structural parameters. As a whole, the results obtained allow us to conclude that the dominating role in the realization of ferromagnetic exchange interactions in the complexes with axial bonds Cu(II)...L (or L...Cu(II)...L) belongs to the delocalization mechanism.

In order to determine the geometrical conformations which are most favourable for the realization of strong ferromagnetic interactions we have calculated the values of J_{dir} and J_{del} as the functions of the geometrical parameters R, ϑ, ψ, and φ ($\alpha = \beta = 0°$) for the axially coordinated complexes Cu(II)...L (models A and C). Analysing the obtained dependences of $J_{dir}(R, \vartheta, \psi, \varphi)$ and $J_{del}(R, \vartheta, \psi, \varphi)$ it was shown that even the considerable variations in the structure of the magnetic fragment Cu(II)...O–N< ($R = 2 \div 4$ Å, $\vartheta = 0° - 90°$, $\psi = 0° - 90°$, $\varphi = 0° - 360°$) the contribution of J_{del} to the exchange interaction remains predominant (approximately an order of magnitude larger than that of J_{dir}). In addition, the ferromagnetic properties of the complex become stronger with decreasing distance R and increasing angles ϑ and ψ. At the same time, the value of the exchange interaction is practically independent of the angle φ of radical L rotation around the bond Cu(II)–O(L). These peculiarities of the conformation dependence of the exchange interaction in such complexes (according to relation (4)) can fully be explained by the proportionality $J_{del} \sim \rho_d(R, \vartheta, \psi, \varphi)$. As an illustration Figure 2 (within the logarithmic

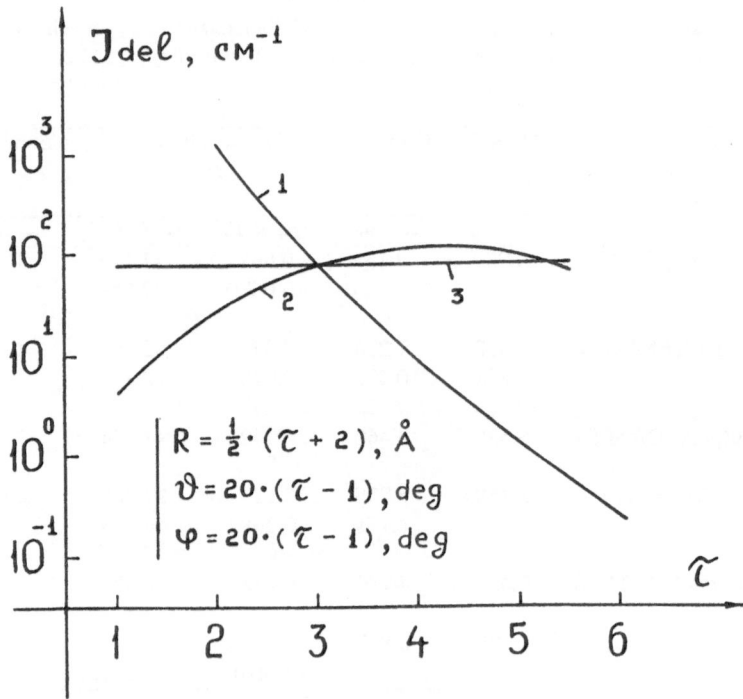

Fig. 2. The dependence of the value of the exchange parameter J_{del} on the geometrical parameters R, ϑ, φ, ($\psi = 60°$) of the axially coordinated complex Cu(II)...L (model C): $1 - J_{del}$ (R) (with $\vartheta = 40°$, $\varphi = 40°$); $2 - J_{del}$ (ϑ) (with $R = 2.5$ Å, $\varphi = 40°$) and $3 - J_{del}$ (φ) (with $R = 2.5$ Å, $\vartheta = 40°$). The values for J_{del} are given on a logarithmical scale.

scale) gives the most typical dependences J_{del} (R), J_{del} (ϑ), and $J_{del}(\varphi)$ calculated for a fixed angle $\psi = 60°$. Thus, the strongest ferromagnetic exchange interactions can be expected for such geometrical conformations of complexes which satisfy the following conditions: the distance R must be minimal (≤ 2.3 Å), angle ϑ must be within the range $60° \div 90°$, angle $\psi \simeq 90°$, and angle φ can have any value. Among the complexes studied, the structure of complex VI is the closest to the optimal one. The greatest value $J_{exp} = 65$ cm^{-1} was obtained just for this complex.

References

[1] A.L. Buchachenko: *Ups. Khim.* **59** (1990) 529.
[2] V.I. Ovcharenko, A.B. Gelman and V.N. Ikorsky: *Zn. Strukt. Khim.* **30** (1989) 142.
[3] A. Ganeschi, D. Gatteschi, A. Grand, J. Laugier, L. Pardi and P. Rey: *Inorg. Chem.* **27** (1988) 1031; and references therein.
[4] D. Gatteschi, J. Laugier, P. Rey and C. Zanchini: *Inorg. Chem.* **26** (1987) 938.
[5] M.H. Dickman and R.J. Doedens: *Inorg. Chem.* **20** (1981) 2677.
[6] O.P. Anderson and C.T. Kuechler: *Inorg. Chem.* **19** (1980) 1417; A. Bencini, C. Benelli, D. Gatteschi and C. Zanchini: *J. Amer. Chem. Soc.* **106** (1984) 5813.
[7] A. Grand, P. Rey and R. Subra: *Inorg. Chem.* **22** (1983) 391.

[8] C. Bernelli, D. Gatteschi, C. Zanchini, J.M. Latour and P. Rey: *Inorg. Chem.* **25** (1986) 4242.
[9] A. Ganeschi, D. Gatteschi, J. Laugier and P. Rey: *J. Amer. Chem. Soc.* **109** (1987) 2191; J.C.S. Farad: *Trans. I* **83 (12)** (1987) 3603.
[10] J.S. Binkley, A.R. Whiteside, R. Krishnan, R. Seeger, D.J. DeFrees, H.B. Schlegei and J.A. Pople: Program N 406, QCPE, University of Indiana, Bloomington, Ind.
[11] D.K. Danovich and B.N. Plakhutin, Program N 57, *Inform. Mater. SFQP*, Siberian Branch of the Academy of Sciences of the USSR, **3** (1989) 24.
[12] P.V. Schastnev and K.M. Salikhov, *Theor. Eksper. Khim.* **9** (1973) 291.

THEORETICAL ASPECTS OF ELECTROACTIVE POLYMERS

V. MUJICA
Escuela de Química
Facultad de Ciencias
Universidad Central de Venezuela (UCV)
Apartado 47102
Caracas 1040
Venezuela

ABSTRACT. Some physicochemical aspects of electroactive polymers, both electronic and ionic conductors are considered. Special emphasis is placed on the discussion of models based on effective Hamiltonians and the relative role of electron-electron interaction, electron-phonon interaction and electronic correlation for explaining the optical gap and local spin densities in conjugated polymers. The nature of the charge carriers and the different current models for conductivity and their ability for explaining the most conspicuous experimental evidences about this kind of materials is also considered.

1. Introduction

The study of electroactive polymers has experienced an explosive growth in the last fifteen years within the area of the so-called "new materials". This development includes the characterization of new compounds with novel properties, particularly electric and catalytic, as well as a better understanding of the basic aspects of the conductivity in such materials. The driving force for this development was the potential technological value of these materials and although initial expectations in this matter were probably too high it is undeniable that considerable new knowledge has been gained and that new and interesting applications have been developed. From the purely academic viewpoint, electroactive polymers form a fascinating family of chemical systems, which exhibit remarkable properties which deserves study on its own right [1–4].

Electroactive polymers comprehend two different types of materials: polymeric electrolytes and organic polymers. These two families of compounds differ in many respects and I will devote two sections to describe their most relevant features separately.

1.1. POLYMERIC ELECTROLYTES

These compounds are included in the class of materials denominated solid electrolytes. They are complexes of inorganic salts with polymers containing a large number of highly polar groups. Structurally they may be partially crystalline as $LiSCN.PEO$ and $NaCF_3SO_3PEI$, or amorphous as $LiSCN.MEEP$ and $NaCF_3SO_3PPO$ (PEI = poly(ethylenimine),

L. A. Montero and Y. G. Smeyers (eds.), Trends in Applied Theoretical Chemistry, 175–185, 1992.
© 1992 *Kluwer Academic Publishers.*

PEO = poly(ethylen oxide), PPO = poly(propylen oxide), MEEP = $-(N = P(OC_2H_4OC_2H_4OCH_3)_2)_n)$.

These are solid materials which can be prepared by forming a suspension of the salt and the polymer in a common solvent, thereafter elimination of the solvent renders the solid complex. The presence of polar groups in the polymer chain is important as a criterion for complex formation since the cohesive energy of the salt lattice must be overcome in the process, this is illustrated by the fact that LiF does not form a complex with PEO while LiCl does, the cohesive energy of the former being 1036 KJ/Mol and that of the latter 853 KJ/Mol [4].

Polymeric electrolytes exhibit both ionic and electronic conduction and there are evidences that the motion of the polymeric chain is strongly coupled to that of the ions. Conduction seems to take place in the amorphous phase as it is evidenced by the fact that there is a sharp decrease in conductivity at the temperature of glass transition. This behavior is in strong contrast with that of fixed-framework inorganic electrolytes of ceramic type where conductivity shows a linear dependence with inverse temperature typical of activated processes [4].

The coupling between the ionic motion and the flexional dynamics of the chain and the fact that conduction takes place in the amorphous phase are two of the most interesting features of these materials. Its proper treatment requires a rather complete knowledge of the phase diagram and structure. From the theoretical point of view, the description of the ionic motion is a difficult problem for many of the ordinary quantum-chemical approximations for molecules and solids, like the separation of the nuclear and electronic degrees of freedom, do not work. In this area one has to resort to approximate treatments like molecular dynamics, Monte Carlo simulations and the use of percolation models [5–7].

Experimentally, the important quantities to be measured are the total conductivity and the fraction of it attributable to the different charge carriers. Such measures are not simple because there is a high resistance to ion flux in the interface electrode-electrolyte [4].

1.2. CONDUCTING ORGANIC POLYMERS

These are highly conjugated organic compounds which undoped are insulators but upon doping acquire semiconductor or even quasi metallic character. Examples of these materials are Poly(p-phenylen) (PPP), poly(p-phenylen sulfide) (PPS) and poly(acetylene) (PA). Conductivity in PA ranges between 10–8 and 102 with doping with AsF_5 with variations of 18 and 16 orders of magnitude for the same quantity for PPS and PPP [1–3].

The transition from insulator to semiconductor and metal in these compounds exhibits some other peculiarities: the optical absorption spectrum shows the appearance of another band as the concentration of dopant is increased and in PPA the conductivity is increasing in a range where the magnetic susceptibility is still constant, showing that the charge carriers are spinless. The phenomenon in the optical absorption spectrum has to do with the appearance of new "mid-gap" states as a consequence of the strong electron-phonon coupling and it stands in sharp contrast with the behavior of ordinary semiconductors for which the effect of the impurity is not to add new states but rather to increase the concentration of charge carriers, electrons or holes. The fact that the magnetic susceptibility is not increasing while the conductivity does, has been explained by invoking the existence of novel spinless charge carriers excitations like in solitons or polaron- or bipolaron-based mechanisms for conduction [8–17].

Another interesting feature of the organic polymers is the highly anisotropic character of the conduction which makes them quasi-linear conductors, as evidenced in the results of nuclear magnetic resonance experiments. The theoretical understanding of all the properties of these systems is still in its beginnings, although some very impressive advances have been made particularly regarding their electronic structure. Much remains to be done in clarifying the dynamics of charge transport.

Present and potential applications of electroactive polymers are numerous: batteries, electrodes, photoelectrical and solar cells, electronic devices and semiconductors with tailored properties to name the most important [18]. This is a typical field where academic and industrial objectives should and can easily converge.

The rest of this article will be devoted to discuss exclusively different aspects models for conjugated organic polymers. This article does not pretend to be neither extensive nor complete. My goal is rather to review part of the enormous literature on this type of material and discuss some of the interplay between different theoretical descriptions and some experimental evidence.

2. Conductivity in Conjugated Organic Polymers

In order to simplify the discussion about this type of materials we shall regard PA as a prototype and much of the content in this section can be adapted to treat other polymers with high degree of conjugation.

I will address three fundamental properties of PA
– Optical gap and midgap absorption.
– Nature of the charge carriers.
– Local spin densities.

These three properties have been selected for they illustrate the interplay of three different factors, namely
– Electron-electron interaction.
– Electron-phonon interaction.
– Electronic correlation.

2.1. GENERAL CONSIDERATIONS

One-dimensional models of various phenomena have long been considered by theoreticians. They are interesting both from the purely formal viewpoint of mathematical physics but also as a test for different approaches. In general they are much simpler than their three-dimensional counterparts, but unfortunately it is also true that in many cases they differ quite drastically in behavior [19, 20].

There are nevertheless certain materials for which one-dimensional models bear a strong resemblance to the actual physical situation. A variety of organic polymers, of which PA is the prototype, are examples of them. Neutral polyenes, $C_{2N}H_{2N}$ or radicals $C_{2N+1}H_{2N+3}$, had long been treated in the quantum chemical literature [21–23], but the revival of both theoretical and experimental interest in them, is due to the fact that PA was the first organic polymer which showed an enormous increase of conductivity (about twelve orders of magnitude) upon doping [3].

An ordinary semiconductor behaves similarly in all three spatial dimensions, PA, in contrast, is one of a type of material which are highly anisotropic and their properties as measured along one axis differ substantially from those measured along other directions [24].

Most of the model Hamiltonians used to describe conjugated hydrocarbons and especially polyenes, share two main ingredients: $\sigma - \pi$ separability is assumed and the nuclei are treated as quasi-classical entities whose motion is coupled by a quadratic elastic potential. The first assumption means that only the π-electrons are explicitly taken into account and that their motion takes place in the mean field created by the nuclei and the σ-electrons. The coupling of the electrons to the lattice vibrations (electron-phonon interaction), as well as the mean field description, are characterized by a number of parameters that appear in the Hamiltonian and which depend on the interatomic distances. Electron-electron interaction is also taken into account approximately by neglecting all the two-electron integrals except the Coulomb integrals. In addition, an orthogonal basis for the on-electron space is ordinarily constructed by taking two spin orbitals (one valence p-state) per carbon atom.

A general enough Hamiltonian that takes into account all the above mentioned is [8, 25]:

$$H = \sum_i \frac{P_i^2}{2M_i} + \sum_i \frac{1}{2} K(u_i - u_{i+1})^2 + H_{\text{PPP}} , \qquad (1)$$

where P_i and M_i are the classical momentum and mass of carbon atom i-th, K is a spring constant, u_i is the displacement of nucleus i-th along the z-axis measured with respect to the equilibrium position and H_{PPP} is the Pariser-Parr-Pople electronic Hamiltonian of the chemical literature [26]. It is given in its second-quantized form by [27]:

$$H_{\text{PPP}} = \sum_i \alpha_i n_{i\nu} + \sum_{ij}{}' \beta_{rs} a_{i\nu}^\dagger a_{j\nu} + \sum_{ij}{}' \frac{1}{2} \gamma_{ij} n_{i\nu} n_{j\nu} , \qquad (2)$$

where the one-electron integrals have been separated in two groups (α_i and β_{ij}), $a_{i\nu}^\dagger$ and $a_{i\nu}$ are creation and annihilation fermion operators for electrons in orbital i with spin ν, primes in the summation signs indicate the omission of terms where the operators correspond to the same spinorbital and $n_{i\nu}$ is the number operator given by

$$n_{i\nu} = a_{i\nu}^\dagger a_{i\nu} . \qquad (3)$$

In principle, all the parameters in the electronic Hamiltonian, α_i, β_{ij} and γ_{ij}, depend on the nuclear positions, hence containing the electron-phonon coupling. In the Born-Oppenheimer approximation one considers fixed successive nuclear positions and this coupling is neglected.

Starting from (2) one can see the relation existing between different model Hamiltonians currently used in molecular and solid state physics. Setting the third term in Equation (2) equal to zero renders the Hückel Hamiltonian or corresponds to what is called the tight-binding approximation with parameter t_{ij} in solid state physics, the correspondence between the parameters being

$$t_{ij} = \sqrt{2}\, \beta_{ij} \qquad (4)$$

[21, 28].

Hubbard Hamiltonian may be obtained from (2) by setting

$$\gamma_{ij} = U \, \delta_{ij} \, , \tag{5}$$

where δ_{ij} is the Kronecker delta symbol. One also refers to the weakly or strongly correlated limits of the PPP Hamiltonian. The former is again the Hückel case, where all two-electron integrals $\gamma_{ij} = 0$, and the latter corresponds to putting the hopping integrals $\beta_{ij} = 0$. These two limiting cases can be exactly solved in molecular orbital (MO) or valence bond (VB) theories.

Starting with the strongly correlated case, where the energy is independent of the spin state, the Hückel term can be introduced as a perturbation (in the limit of $\gamma_{ii} \gg \beta_{ij}$) that lifts the spin degeneracy. Using standard perturbation theory, it is possible to prove that the first order correction vanishes. The second order correction, H_S, for a chain of identical atoms, can be written in terms of the spin operators per site

$$\mathbf{s}_i = (s_i^+, s_i^-, s_i^z)$$

as a spin-dependent exchange term, i.e., [29]

$$H_S = \sum_{rs} J_{rs} \mathbf{s}_r \cdot \mathbf{s}_s \, , \tag{6}$$

where

$$J_{rs} = \frac{2 \, |\beta_{rs}|^2}{\gamma_{rr} - \gamma_{rs}} \, . \tag{7}$$

These results are obtained provided one remains in one of the eigenspaces of the total number operator per site

$$N_r = n_{r+} + n_{r-}$$

which are labeled by the eigenvalues 0, 1, 2 that correspond to the different occupations of the p orbital. In practice it means that one can use (7) for studying the low-lying electronic excitation spectrum provided one does not change the occupation per site.

The operator (6) belongs to a family of widely used effective operators, known as Heisenberg, exchange or spin operators [30, 28]. The parameters of the Heisenberg Hamiltonian are the J_{rs}'s and depending on their sign one speaks of a Heisenberg ferromagnet ($J > 0$) or antiferromagnet ($J < 0$). To the level of approximation considered, the low-lying states of the electronic Hamiltonian could then be approached as those of an antiferromagnet.

To end this general considerations, I should add a word about electronic correlation. Whereas electron-electron and electron-phonon interaction must be incorporated in the model Hamiltonian [25, 31], electron correlation has to be included in the model wavefunction as well, provided the Hamiltonian allows for electronic interaction. Thus Hartree-Fock level description allows for electronic interaction but a single-determinant wavefunction does not take into account coulombic electronic correlation although it does incorporate Pauli spin correlation to a certain extent. On the other hand Hückel description is both a non-interacting and uncorrelated description.

2.2. OPTICAL GAP, MIDGAP ABSORPTION, NATURE OF THE CHARGE CARRIERS AND SPIN DENSITIES

The photoabsorption spectrum of PA has a gap, corresponding to a singlet-singlet transition, of about 1.5 eV. The existence of this energy gap between the conduction and valence band of PA is responsible for it being a semiconductor. Upon doping with either p- or n-type dopants midgap absorption is observed. The undoped material has a concentration of free radicals of about 1:3000 which determines a non vanishing spin density in the carbon atoms which has been experimentally detected to have positive and negative alternating values in odd chains that correspond to the free radicals [32]. For low doping levels the conductivity increases while the Pauli magnetic susceptibility remains constant. A brief consideration of the way these observations have been accounted for, based on the model operator (1), is very instructive.

For a uniform chain of n sites (equal bond lengths) of period L and at the Hückel MO level one easily obtains the result that the energy gap between the lowest empty (LUMO) and highest filled (HOMO) (or equivalently, between the valence and conduction bands) is given by [21, 33]

$$\Delta_{n,L} = -4\beta \sin \frac{\pi}{2(2n+1)} ,$$ (8)

where β is the single resonance integral corresponding to all-equal bonds. By taking the appropriated limit one sees that for large n, the gap vanishes. The same result is obtained at the restricted Hartree-Fock level [23], indicating that electron interaction without correlation is not able to explain the existence of the gap.

On the other hand, if one considers, as Kuhn did many years ago [34] that bond alternation should persist in long polyenes then Hückel MO theory for a chain with a period of $2L$ gives for the gap the result

$$\Delta_{n,2L} = 2 \sqrt{\beta_1^2 + \beta_2^2 + 2\beta_1\beta_2 \cos \frac{2\pi n}{2n+1}} ,$$ (9)

where now two different β's have been used to take into account the alternating single and double bonds. It is easily seen that for large n one obtains a finite gap whose magnitude is

$$\Delta = \lim_{n \to \infty} \Delta_{n,2L} = 2 |\beta_1 - \beta_2| .$$ (10)

For n even there are not states in the middle of the gap, for n odd (which would correspond to a free radical) there is a state in the middle of the gap, the magnitude of the energy gap still being given by (10).

A conclusive experimental evidence for bond alternation in PA remained elusive for many years and it was only recently that x-ray and NMR measurements have definitely settled the issue and showed that the double bond is about 0.03 Å shorter [35]. However, the question remains as to what is the physical origin of bond alternation.

Part of the answer is an effect known in the solid state literature as Peierls transition in one dimensional metals, which in fact is a demonstration of why one-dimensional conductors can not be found [36]. Peierls showed that at low temperature a distortion of a one-dimensional lattice that doubles its period leads to a small gap opening at the Fermi energy. The decrease in electronic energy due to states being lowered to form the gap makes the distorted state the stable one, despite the cost in lattice energy. This effect might also be considered as an example of Jahn-Teller distortions which are well known in the chemical literature. In fact, Longuet-Higgins and Salem [37] proved within the context of LCAO theories that the infinite polyene is unstable with respect to dimerization. This is in agreement with the fact that the topology of the carbon backbone dominates the spectra of alternant hydrocarbons like polyenes, as was first noted by McLachlan [38]. This topological dominance is the reason behind the comparability of Hückel, PPP and even free-electron results for these systems.

In modern language, Peierls transition is seen as the result of electron-phonon interaction [39]. The acoustic phonons (the quanta of lattice vibrations), although having very little energy compared to that of the electrons may have comparable momentum and can scatter electrons from the boundaries of the Fermi surface, which for a one-dimensional material consists only of two points at $\pm k_F$. This gives rise to a strong interaction of the electrons with acoustic phonons of frequency $2k_F$. Fröhlich's treatment shows that the gap increases with the strength of the electron-phonon coupling and there exists a temperature T_P above which there is not energy gain with the distortion of the lattice and the material becomes a conductor. This behavior has been observed in several organic polymers [24].

The other contribution to the energy gap comes from electronic correlation. Since the work of Ovchinikov it is clear that even for a uniform chain if one includes electron correlation an energy gap is obtained [22]. It is also clear that the magnitude of the dimerization, although it explained qualitatively the appearance of the gap, was not enough to give a quantitative agreement with the observed value. In addition, bond alternation has also been proved to be induced by correlation effects for a Hubbard-type interaction [40].

In summary one can conclude that bond alternation and the existence of the gap are intimately related and that both electron-phonon interaction and electronic correlation are needed in order to achieve a proper description.

In conventional semiconductors donor (acceptor) doping results in generation of localized states near the conduction (valence) band. These states and consequently the peaks in the absorption spectrum depend on the dopant [28]. Instead, in PA there seems to be states which are roughly localized in the middle of the gap and the shape of the spectrum, except for the intensity, is rather independent of the dopant nature and concentration, for moderate doping. These observations, together with the fact that conductivity increases without appreciable change in the Pauli magnetic susceptibility, suggesting that spinless carriers are responsible for the conduction at low doping, were first rationalized in terms of a model that entirely neglects electron interaction but that takes into account electron-phonon interaction. The Sue-Schrieffer-Heeger (SSH) model Hamiltonian is given by [8]

$$
\begin{aligned}
H_{SSH} = &\sum_n \frac{P_n^2}{2M} + \sum_n \frac{1}{2} K(u_{n+1} - u_n)^2 \\
&- \sum_{n\sigma} \beta_{n+1,n}(a_{n+1,\sigma}^\dagger a_{n,\sigma} + a_{n,\sigma}^\dagger a_{n+1,\sigma})
\end{aligned}
\tag{11}
$$

the electron-phonon coupling is introduced through a linear coupling term

$$
\beta_{n+1,n} = \beta_0 - \alpha(u_{n+1} - u_n),
\tag{12}
$$

where α and β_0 are parameters of the model. It is obvious that the SSH model operator is derived from (1) by considering only nearest-neighbors hopping interaction and neglecting the two-electron part of the PPP Hamiltonian.

By considering first only the static part of (11), that is

$$H_{\text{SSH}}(u) = H_{\text{e-ph}} + 2NKu^2 \,, \tag{13}$$

where $H_{\text{e-ph}}$ corresponds to the third term in (11) giving the electron-phonon interaction, it was possible to solve for the ground state of (13) [8] as an expansion in the lattice deformation

$$\frac{E_0(u)}{N} \approx -\frac{4\beta_0}{\pi} - \frac{2\beta_0}{\pi} \left[\ln\left(\frac{4}{z}\right) - \frac{1}{z} \right] z^2 + \frac{K\beta_0^2 z^2}{2\alpha^2} + \ldots \tag{14a}$$

$$z = \frac{2\alpha u}{\beta_0} \,. \tag{14b}$$

Equation (14a) leads to the conclusion that there is a degeneracy in the ground state energy corresponding to two equivalent distortions of the uniform infinite lattice but with different sign. This is tantamount to assigning a "direction" to the double bonds: starting from left to right (A phase) or from right to left (B phase). It has already been shown that the energy lowering caused by bond alternation is enough to explain the band gap qualitatively. Using a value of 1.4 eV for the gap and within the SSH model, a value of 0.04 Å was obtained for the distortion of the bond.

The type of ground state degeneracy alluded to above implies that the system supports non-linear excitations known as topological solitons, i.e., domain walls separating regions having different ground states. The formation of a soliton brings about the formation of a midgap state which in a neutral odd chain has single electronic occupancy, hence corresponding to a situation where there is zero charge and spin 1/2. Upon doping there would be exchange of one electron with the dopant, therefore leading to situations where the midgap state would be doubly populated or empty, depending on the direction of the electronic transfer, having zero spin in both cases. In this way the SSH explained the inverse charge-spin relation needed to account for the magnetic susceptibility results. In a way the electronic consequences of soliton formation are the ones one could have obtained from the simple Hückel model for an odd alternating chain. However, important additional insight is provided by the soliton model. The energy of soliton formation was calculated to be smaller than the band gap, hence making them more favorable elementary excitations than promoting an electron to the conduction band. This also implies that midgap absorption was rather a property of the chain, intimately related to its quasi one-dimensional character, and that the role of the dopant, at least for low doping level, was restricted to provide an electron (hole). They are not point disruptions and by a variational procedure its length was estimated to be about 15 sites [8]. In the SSH model, the solitons are free to travel along the chain and the spinless charge carriers are associated with them.

The success of the soliton model led to a large number of articles trying to test the validity of the proposed conduction mechanism and also assess the influence of electronic correlation in its predictions, both using conventional molecular quantum-chemical methods and also band structure calculations [31, 41–45]. From the quantum chemist's viewpoint it was very important to settle down the question of the importance of electronic correlation for it was well known that energy ordering in finite polyenes was inconsistent with any one-electron model [46].

Agreement does not exist as to the role of solitons in charge transport, which has been questioned by a number of authors. The main objections have to do with the inclusion of inter-chain transport and interactions. The latter causes three-dimensional pinning of solitons and the former is increasingly considered to be of importance [47–49].

Other type of elementary excitations, polarons and bipolarons, are also supported by systems like PA. They might be considered as interacting solitons or also created by injection of a single electron or hole. I will not discuss them in any detail but only mention that they can also be invoked to explain the inverse spin-charge relationship in the low doping regime. They are extremely important for understanding conductivity in systems with non-degenerated ground states yet having a large degree of conjugation for these systems do not support topological solitons of the kind described for PA [1, 43, 49].

Local spin densities in PA is the last aspect I will consider of this system. They are wrongly predicted in the SSH model or in any one-electron model, including both restricted and unrestricted Hartree-Fock [32]. Its right description requires the introduction of correlation and it can be achieved both including it as a correction to the unrestricted Hartree-Fock result [1, 50] or using a Heisenberg Hamiltonian of the type in Equation (6), both in Monte Carlo simulations [25, 51] or in direct calculation using approximate ground state wavefunctions [8].

There is an interesting connection between a continuum stochastic model of domain wall motion [17] which would account for the displacement along the PA chain and the use of Heisenberg-type Hamiltonians. In the stochastic model an elastic potential composed of a quadratic plus a quartic term is used for the motion of the nuclei. Such a potential also provides the framework for the existence of domain walls (topological solitons) for it has two degenerated static ground states. From the point of view of the quantum mechanical description this effective potential should arise as a combination of the quadratic part in (1) plus an effective quartic term coming from the electronic contribution averaged over the electronic ground state, that is, within the approximations involved in the use of Heisenberg operators, the effective potential $V(u)$ for the average nuclear displacement should be given by

$$V(u) = Au^2 + Bu^2 = 2NKu^2 + \sum_i J_{i,i+1} \langle s_i \cdot s_{i+1} \rangle , \qquad (15)$$

where $\langle .. \rangle$ stands for the average over the electronic ground state. By using Equations (7) and (12) and the Mataga-Nishimoto parametrization for the electron integrals:

$$\gamma_{ij} = (14.397 \, \text{eV})/[(14.397 \, \text{eV})/\gamma_{ii} + R_{ij}] , \qquad (16)$$

where R_{ij} is the internuclear distance, one can indeed obtain (15) as the leading terms in an expansion in the displacement u [52].

There are still many aspects of the behavior of conducting polymers which remain to be clarified. The controversy as to the role of the different type on non-linear excitations in the conduction mechanism is still alive. For recent work in the field I submit the reader to [53–56].

References

[1] D.S. Boudreaux: *Proceedings of the Advanced Summer School on the Electronic Structure of New Materials*, H. Stubb (ed.), Rep. No. 40, The Swedish Academy of Engineering Science in Finland (1985).

[2] S. Stafström: 'Conducting Polymers, a Theoretical Study of Disorder Effects in a Model Polymer and Defect States in Polyacetylene', Linköping Studies in Science and Technology, Dissertation No. 125, Linköping (1985) and references therein.

[3] S. Etemad, A.J. Heeg:r and A.G. MacDiarmid: *Ann. Rev. Phys. Chem.* **33** (1982) 443.

[4] M.A. Ratner and D.F. Shriver: *Chem. Rev.* **88** (1988) 109.

[5] R. Granek, A. Nitzan, S.D. Druger and M.A. Ratner: 'Dynamics of Ionic Motion in Polymeric Ionic Conductors', unpublished.

[6] S.D. Druger, M.A. Ratner and A. Nitzan: *Phys. Rev.* **B 31** (1985) 3939.

[7] S.D. Druger, A. Nitzan and M.A. Ratner: *J. Chem. Phys.* **79** (1983) 3133.
     S.H. Jacobson, M.A. Ratner and A. Nitzan: *J. Chem. Phys.* **77** (1982) 5752.

[8] W.P. Su, J.R. Schrieffer and A.J. Heeger: *Phys. Rev.* **B 22** (1980) 2099.
     V. Mujica, N. Correia and O. Goscinski: *Phys. Rev.* **B 32** (1985) 4178, ibid 4186.

[9] M. Takahashi and J. Paldus: *Can. J. Phys.* **62** (1984) 1226.

[10] A.J. Heeger, S. Kivelson, J.R. Schrieffer and W.P. Su: *Rev. Mod. Phys.* **60** (1989) 2099.

[11] A.S. Davydov: *Sov. Phys. Usp.* **25** (1982) 898.

[12] R.R. Chance, D.S. Boudreaux, J.L. Brédas and R. Silbey: *Phys. Rev.* **B 27** (1983) 1440.

[13] J.L. Brédas, J.C. Scott, K. Yakushi and G.B. Street: *Phys. Rev.* **B 30** (1984) 1023.

[14] D.S. Boudreaux, R.R. Chance, J.L. Brédas and R. Silbey: *Phys. Rev.* **B** (1983) 6927.

[15] K. Yakushi, L.J. Lauchlan, G.B. Street and J.L. Brédas: *J. Chem. Phys* **81** (1984) 4133.

[16] A.H. Luther: *Solitons*, R.K. Bullough and P.J. Caudrey (eds.), Springer, Berlin (1980).

[17] Y. Wada and J.R. Schrieffer: *Phys. Rev.* **B 18** (1978) 3897.

[18] D. Bloor: *Nature* **335** (1988) 115.
     R.G. Linford (ed.): *Electrochemical Science and Technology of Polymers*, Elsevier, London (1987).

[19] E. Lieb and D.C. Mattis: *Mathematical Physics in One Dimension*, Acad. Press, N.Y. (1966).

[20] H.C. Fogedby: *Theoretical Aspects of Mainly Low Dimensional Magnetic Systems*, Springer-Verlag, Berlin (1980).

[21] C.A. Coulson: *Proc. Roy. Soc. London, Ser. A* **164** (1938) 383; ibid **A 169** (1939) 413.
     H.C. Longuet-Higgins and L. Salem: *Proc. Roy. Soc. London, Ser. A* **251** (1959) 172.
     L. Salem: *Molecular Orbital Theories of Conjugated Systems*, Benjamin, New York (1966).
     J.E. Lennard-Jones: *Proc. Roy. Soc. London, Ser. A* **158** (1937) 280.

[22] A.A. Ovchinikov, I.I. Ukrainskii and G.V. Kventzel: *Sov. Phys. Usp.* **15** (1973) 575. (*Ups. Fiz. Nauk.* **108** (1973) 81.)

[23] J. Paldus and E. Chin: *Int. J. Quantum Chem.* **24** (1983) 373.

[24] E.M. Conwell: *Physics Today* (1985) June.

[25] D.K. Campbell, T.A. De Grand and S. Mazumdar: *Phys. Rev. Lett.* **52** (1984) 1717.

[26] R. Pariserand and R.G. Parr: *J. Chem. Phys.* **21** (1953) 767.

[27] J. Linderberg and Y. Öhm: *Propagators in Quantum Chemistry*, Academic Press, N.Y. (1973).

[28] O. Madelung: *Introduction to Solid State Physics*, Springer-Verlag, Berlin (1978).

[29] L.N. Bulaevskii: *Sov. Phys. JEPT* **24** (1967) 154. (*Zh. Eksp. Teor. Fiz.* **51** (1966) 230.)

[30] A.I. Akhiezer, V.G. Bar'Yakhtar and S.V. Peletminskii: *Spin Waves*, North-Holland, Amsterdam (1968).

[31] Z.G. Soos and L.R. Ducasse: *J. Chem. Phys.* **78** (1983) 4092.

[32] J.E. Hirsch and M. Grabowski: *Phys. Rev. Lett.* **52** (1984) 1713.

[33] R. Hoffmann: *Solids and Surfaces*, VCH Publishers, Inc., N.Y. (1988).

[34] H. Kuhn: *Helv. Chim. Acta* **31** (1948) 1441.

[35] C.R. Fincher, D. Moses, A.J. Heeger, A.G. MacDiarmid and J.B. Hastings: *Phys. Rev. Lett.* **48** (1982) 100.
     C.S. Yannoni and T.C. Clarke: *Phys. Rev. Lett.* **51** (1983) 1191.

[36] R. Peierls: *Quantum Theory of Solids*, Clarendon Press, Oxford (1955).

[37] H.C. Longuet-Higgins and L. Salem: *Proc. Roy. Soc. London, Ser. A* **252** (1959) 172.

[38] A.D. McLachlan: *Mol. Phys.* **2** (1959) 276.

[39] H. Fröhlich: Proc. Roy. Soc. London, *Ser. A* **223** (1954) 296.

[40] P. Horsch: *Phys. Rev.* **B 24** (1981) 7351.

[41] I.R. Ducasse, T.E. Miller and Z.G. Soos: *J. Chem. Phys.* **76** (1982) 4094.

[42] D.S. Boudreaux, R.R. Chance, J.L. Brédas and R. Silbey: *Phys. Rev.* **B 28** (1983) 6927.

[43] S. Stafström and K.A. Chao: *Phys. Rev.* **B 29** (1984) 2255.

[44] J. von Boehm, P. Kuivalainen and J.L. Calais: *Solid State Comm.* **48** (1983) 1085.

[45] H. Fukutome and M. Sasai: *Prog. Theor. Phys.* **69** (1983) 1.

[46] K. Schulten, I. Omine and M. Karplus: *J. Chem. Phys.* **64** (1976) 4422.

[47] P.L. Danielsen and R.C. Ball: *J. Physique* **46** (1985) 131.

[48] T. Yamabe, K. Tanaka, S. Yamanaka, T. Koike and K. Fukui: *J. Chem. Phys.* **82** (1985) 5737.
[49] J.L. Brédas, R.R. Chance, R. Silbey, G. Nicolas and P.H. Durand: *J. Chem. Phys.* **75** (1981) 255.
[50] M. Sasai: *Int. J. Quantum Chem.* **37** (1990) 559.
[51] D.K. Campbell, T.A. De Grand and S. Mazumdar: *J. Stat. Phys.* **43** (1986) 803.
[52] V. Mujica and O. Goscinski: unpublished.
[53] C. Yang, W. You-Liang and C. Bo: *Int. J. Quantum Chem.* **37** (1990) 679.
[54] H. Kamimura (ed.): *Theoretical Aspects of Band Structure and Electronic Properties of Pseudo-One-Dimensional Solids*, D. Reidel, Dordrecht (1985).
[55] *Solitons and Polarons in Conducting Polymers*, World Scientific, Singapore (1988).
[56] Y. Takada and M. Kohmoto: *Phys. Rev.* **B 41** (1990) 8872.

THEORETICAL APPROACHES TO THE VIBRATIONAL SPECTRA OF POLYATOMIC MOLECULES

JAMES E. BOGGS
Department of Chemistry
The University of Texas
Austin
Texas 78712
U.S.A.

ABSTRACT. Computational chemistry can make useful contributions to other branches of chemistry by supplementing infrared and Raman spectroscopy as a source of information about the mechanical flexibility of molecules. A major problem in doing so is the difficulty in getting sufficient accuracy and reliability in theoretical prediction of vibrational force fields, anharmonicity, and spectral frequencies. Two methods for solving this problem are outlined.

First, advantage can be taken of the remarkable uniformity in the computational error for a given type of vibration (*e.g.*, C—H bond stretch, heavy atom bend, etc.) in different molecular environments. Similarly, the contribution of anharmonicity for a given type of motion is highly transferable. This permits development of a small set of scaling factors which can be used as transferable corrections between known and unknown molecules.

Alternatively, a very high-level quantum chemical computation of the vibrational potential energy hypersurface followed by a variational solution of the vibrational Schrödinger equation can yield highly accurate anharmonic frequency predictions for small molecules. Advantages and limitations of the two procedures are discussed.

Introduction

Before examining the ability of modern quantum chemistry to predict the vibrational properties of polyatomic molecules, it is useful to consider why such an effort should be undertaken. After all, there are perfectly good and relatively simple experimental techniques by which the infrared and Raman spectra of stable molecules can be recorded with acceptable accuracy, so why should they be calculated? The computed frequencies are never obtained with the reliability and accuracy with which they can be measured. If quantum chemistry is to find its role as one member of the wide range of techniques available to the practicing chemist, computation of vibrational properties can only be justified if there are chemical questions for which contributions to the answers come most easily by this approach.

First, it must be realized that the experimental vibrational spectroscopist faces a formidable task. For any but very small molecules, it is impossible to provide an unequivocal assignment of all of the bands observed in the spectrum. The usual approach is to make observations on as many isotopic species as possible, perhaps incorporate any information that can be obtained from other vibrational observables such as Coriolis or

L. A. Montero and Y. G. Smeyers (eds.), Trends in Applied Theoretical Chemistry, 187–197, 1992.
© 1992 *Kluwer Academic Publishers.*

TABLE I
Experimental and computed spectrum of NH$_3$ [a].

Experimental (as measured)	Experimental[b] (harmonic)	Computed[c] 6–31G* MP2
3444	3577	3659
3337	3506	3504
1627	1691	1852
950	1022	1166

[a] Frequencies in cm^{-1}.

[b] J. L. Duncan and I. M. Mills, *Spectrochim. Acta*, **20**, 523 (1964).

[c] W. J. Hehre. L Radom, P. v. R. Schleyer, and J. A. Pople, *Ab Initio Molecular Orbital Theory*, Wiley, New York, 1986, pp. 234–235.

centrifugal distortion constants, and attempt to construct a force field (a matrix of harmonic force constants and interaction constants) for the molecule. Unfortunately, the force field is greatly under-determined by the data except for the smallest molecules. It is common to carry over some force constants or coupling constants from related molecules or even to assume that many of the coupling constants can be ignored. This difficulty provides an area in which computed vibrational frequencies can be of great assistance in the interpretation of experimental spectra. Even an approximate prediction of the frequency corresponding to a given vibrational motion may be sufficient to permit a correct identification of the corresponding observed spectral band. By this means it has been possible, for example, to obtain [1] a complete spectral assignment for γ-pyrrone, $C_4H_4O_2$, by coupling quantum chemical predictions with observations on only one isotopic species.

Aside from providing assistance in the assignment of new spectra, computed vibrational frequencies are of help in completing and correcting previously reported assignments. It is rare that a published spectral assignment for a molecule containing a dozen atoms or more is not found to contain misassignments or at least controversial assignments which can be corrected by spectral prediction from computation.

One very important area for application of computed spectra is in the identification of newly discovered molecules, ions, or radicals, particularly those for which detailed experimental manipulation is difficult or impossible. In this class, one may think of astronomical species, gaseous materials in flames or other high temperature environments, or highly reactive materials observed during chemical reactions.

It is unfortunate that many commonly used computational computer programs concentrate on the prediction of vibrational frequencies to the exclusion of the more fundamental molecular property of vibrational force fields. It is not possible to calculate a force field uniquely from the spectrum, but the reverse calculation is simple, at least at the harmonic oscillator level of approximation. From the force field, one can also calculate other chemically useful properties including Coriolis constants, centrifugal distortion constants used in rotational microwave spectroscopy, and the mean amplitudes of vibration needed for analysis of gas-phase electron diffraction results.

Accurate experimental determination of fundamental thermodynamic properties has become a lost art, replaced by prediction of basic data such as heat capacities, heat of formation, absolute entropy and Gibbs free energy of formation by the methods of statistical mechanics. The vibrational partition function can be obtained only if the vibrational energy levels are known, especially the low-lying ones which are frequently more difficult to obtain by experiment. Computation can make a great contribution here. Even more significantly, overtone, combination and occasionally long sequences of hot bands become important at elevated temperatures. The input for the contribution these make to the partition function can be obtained from calculations that take vibrational anharmonicity into account.

The Problem

The first column of Table I shows the observed fundamental band center frequencies of NH_3. It would be unfair to compare computed frequencies directly with these since it is customary to make computations within the limits of the harmonic oscillator approximation while nature is anharmonic. If sufficient experimental data are available on overtone bands of all of the fundamentals, it is possible to correct the experimental observed frequencies for the effect of vibrational anharmonicity. This has been done to provide the values given in the second column of the table. Note that the effect of anharmonicity is quite large, ranging from 64 cm^{-1} to 169 cm^{-1} for the NH_3 bands. In almost all cases of interest, sufficient experimental information is not available to make the correction for anharmonicity, so an absolutely perfect calculation of the harmonic frequencies would be this far off from reality. Any claim that calculated harmonic frequencies agree with experimental observations to a high level of accuracy must be based on chance cancelation of the error due to anharmonicity by computational error.

The third column of Table I shows the NH_3 fundamental frequencies calculated at a level of approximation high enough that it normally gives excellent agreement between computed and experimental molecular geometries yet not so large that it cannot be used for the larger molecules for which we now desire to calculate vibrational frequencies. The results show large deviations compared with the experimental harmonic frequencies given in column 2, with an average deviation of 97 cm^{-1}, but with individual errors varying from +161 cm^{-1} to −2 cm^{-1}. The results show that a level of computation that gives good results for molecular geometries is not adequate for the computation of even harmonic vibrational frequencies.

It thus appears that there are two major reasons why computed vibrational band frequencies show quite unsatisfactory agreement with experimental ones. First, the computed harmonic frequencies cannot be expected to reproduce real anharmonic ones. Secondly, unless a level of calculation so large that it can only be applied to quite small molecules is used, there are computational errors of large magnitude. Both of these sources of discrepancy may be large. If the computation of vibrational frequencies and other molecular vibrational behavior is to make the contribution to chemistry that is hoped, some means must be found to solve the problem of the large errors shown here. One common approach, computing harmonic frequencies at a very slightly higher level, say 6–31G**+MP2, and relying on the residual computational error from the small basis set, low-level treatment of electron correlation and neglect of anharmonicity to roughly cancel each other so that the directly computed frequencies can be compared with measured ones, sometimes give reasonably good results but is sometimes greatly in error. An approach which does not rely on the chance cancelation of large errors would be preferable.

Solution Number 1: Scaling

A general observation that has often been made, especially for organic compounds, is that force constants computed at a simple double-zeta Hartree-Fock level are systematically too large by 15–20% to reproduce vibrational frequencies in agreement with experiment. A common approach, then, is to compute the force field matrix at this easily affordable level and multiply at least the diagonal elements by 0.8 or some such factor. An equivalent correction can be made by calculating the fundamental frequencies from the directly computed force constants and then multiplying the frequencies by 0.9, since frequencies are related to the square root of force constants.

Use of a constant multiplicative correction factor produces a marked improvement in the average agreement between harmonic frequencies computed with a modest basis set and observed experimental frequencies. It does not, however, guarantee that there will be such an improvement for all vibrational bands or for any specific band which may be of interest. For example, the vibrational force field of pyrrole was calculated [2] at a quantum chemical level such that 0.8 would have been a satisfactory average correction factor for all of the motions. For the N—H wagging motion of this molecule, however, a correction factor of 0.45 was required to achieve similar agreement between experiment and theory. It is also obvious that no single multiplicative factor could bring all of the computed frequencies for NH_3 shown in Table I into uniform agreement with the experimental measurements.

Further progress can be made by approaching the problem from the viewpoint of a chemist. Chemists are accustomed to thinking of molecules as being made up of atoms and groups of atoms that retain some of their characteristics regardless of the molecule in which they are found. The discrepancies noted here between computed and observed vibrational frequencies arise primarily from (1) use of inadequate basis sets, (2) neglect of electron correlation, and (3) neglect of anharmonicity. All of these might be expected to be essentially constant for a similar chemical bond type, and hence for a given vibrational motion, in various molecules. For example, there is no reason to expect that neglecting electron correlation would affect the electron density distribution in a $C\equiv C$ triple bond in the same manner as in a C—C single bond, but the effect might be very similar in all $C\equiv C$ triple bonds regardless of their molecular environment. Similarly, the anharmonicity of a given C—H bond ought to be much like that of any other C—H bond.

Transferable corrections based on these chemical considerations cannot be applied to vibrational frequencies, since these do not in general arise from motions of an individual bond but rather have contributions from motions of a variety of groups in the molecule. Similarly, the corrections cannot be made on force constants expressed, as they sometimes are, in Cartesian coordinates, since there is nothing transferable between molecules about the motion in the x direction of all carbon atoms. They can very usefully be applied, however, to force constants expressed in sets of internal coordinates that have been designed to emphasize this property of transferability. Suggestions for such a coordinate system have been given [3].

The scaled quantum mechanical (SQM) force field method has been fully described previously [3, 4] and has now been applied in numerous published studies. The basic idea is to perform similar calculations on the molecule whose spectrum is to be predicted and on one or more related molecules for which the spectrum and its assignment are known. Care must be taken to use the best known approximation to the true equilibrium geometry for each molecule as the reference geometry around with the vibrational energy expansion is performed, to use similar internal coordinates, and to use the same quantum chemical level of computation. Specifically, to study molecule X, one computes the quadratic force constant matrix F_{th} for a related molecule Y. A matrix of scale factors, C, to correct F_{th} to a matrix F which gives the best least-squares fit to the experimentally known frequencies of molecule Y is obtained:

$$F = C^{1/2} F_{th} C^{1/2}.$$

Next the matrix F_{th} is calculated for molecule X and the set of scale factors C is transferred unchanged to obtain a scaled force constant matrix F for the unknown molecule. It has been found by experience that good results are obtained if the off-diagonal elements of C are not varied independently, but are taken as the geometric mean of the diagonal elements which they connect. Furthermore, common diagonal scale factors can be used for groups of related motions, such as all C—H stretches or all heavy atom out-of-plane torsional motions. Consequently, the vibrations of even large molecules can be described by use of sets of only 5 to 10 separate scaling factors.

There are now very many published examples from numerous laboratories in which small sets of scaling factors have been used. One of the earlier applications can still serve as an illustration. The harmonic force constant matrix of benzene was calculated using a medium-sized (4–21) basis set and no treatment of electron correlation [5] and this matrix was used to predict the spectrum of fundamental vibrational frequencies of benzene. The internal motions were divided into six chemically related groups and the six elements of the C matrix were derived to give the best fit to the known spectrum of benzene. While the scale factors obtained were in the general range of 0.8, they varied from 0.739 to 0.911 for the different types of motions, illustrating the point that there is more similarity within chemically related types of motion than can be made use of by application of a single scaling factor.

The observed vibrational fundamentals of benzene were fit with an average deviation of less that 6 cm^{-1} by the scaling procedure. This is not particularly noteworthy since it represents a simple fitting of known data, but the point is that the scaling factors, C, which have been obtained for benzene are now transferable to similar motions in other molecules. The first example tested was pyridine for which calculations were made [6, 7] in exactly the same manner as had been done for benzene. The directly computed force constant matrix was then scaled using the benzene scaling factors and the fundamental transition frequencies of pyridine calculated. Comparison with the pre-viously observed spectrum of pyridine showed a mean difference of only 5.7 cm^{-1} for the in-plane motions and 8.5 cm^{-1} for the out-of-plane motions, excluding the C—H stretches for which the experimental values are dubious because of probable small Fermi resonances. Including the C—H stretches, the deviations averaged 9.6 cm^{-1} for the in-plane motions.

A similar test was made transferring the benzene scaling factors to correct the computed force constants of naphthalene [8]. The average difference between the scaled computed frequencies and the observed ones was only 6.5 cm^{-1}, excluding the C—H stretches which had not been individually assigned experimentally. The largest single deviation, 21 cm^{-1}, was for a B_{2u} mode which is less sensitive to any diagonal force constant than it is to the coupling constant between modes in the two separate rings, a feature which has no counterpart, of course, in the single ring of benzene.

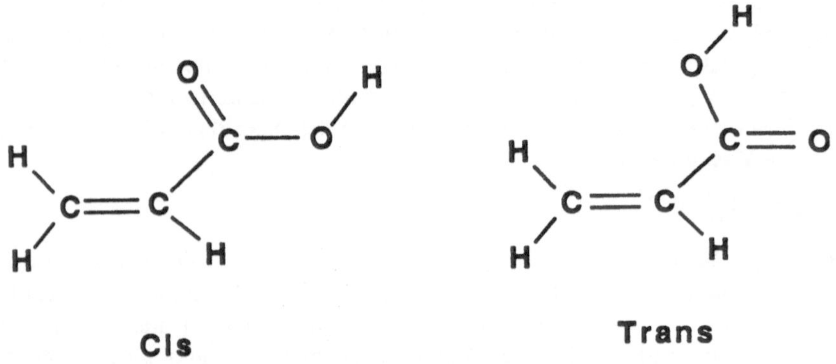

Fig. 1. *Cis-* and *trans-* conformers of acrylic acid.

The same procedure, with scaling factors unchanged from benzene, has been used with great success in a large number of studies, including work on aniline [9], toluene [10], fluorobenzene [11], benzonitrile [12], and phenylacetylene [13]. Furthermore, the identical benzene scaling factors have been applied in work on a large number of less closely related 5- and 6-membered heterocyclic rings, including even borazine [14] in which there are no carbon atoms and the ring consists of alternating boron and nitrogen atoms.

It should be remembered that the scaling procedure, as described here, corrects not only for errors of quantum chemical calculation but also for the effects of vibrational anharmonicity. Thus, even an imaginary calculation with an infinite basis set and a perfect treatment of electron correlation would need to be scaled to agree with the anharmonic vibrations of real molecules. Further, there are possible computational errors for which scaling cannot correct. For example, there might be a low-lying excited electronic state in a molecule (so that a multi-configuration reference state should really have been used) and no corresponding difficulty in the systems chosen for reference. Nevertheless, this technique has shown its ability to predict vibrational spectra accurate within less than 20 cm^{-1} or so for scores of molecules.

One additional study might be mentioned to illustrate the accuracy of the invariance of computational error for similar motions in closely related molecules. The vibrational spectrum of acrylic acid frozen in a matrix showed split bands [15], presumably corresponding to simultaneous spectra from *cis-* and *trans-* forms of the acid (see Figure 1). Irradiation with ultraviolet light made it possible to distinguish two sets of bands, one increasing in intensity and one decreasing after irradiation. Presumably Set 1 and Set 2 corresponded in some order to the *cis-* and *trans-* conformations, but the experiment did not make it possible to determine which was which. Attempts to distinguish between them by a normal coordinate analysis failed because the small differences between the band frequencies demanded a greater accuracy of prediction than could be obtained in this manner.

The two spectra were distinguished using scaled SCF-level calculations of the two forms [16]. The directly computed force fields were not accurate enough to make the distinction, of course, but these were scaled using scaling factors previously optimized [4] to give a simultaneous fit to the observed spectra of ethylene, formaldehyde, glyoxal, acrolein, and

TABLE II

Frequency differences between *cis*- and *trans*-acrylic acids.[a]

Approximate description	Calculated *trans* − *cis*	Observed Set 1 − Set 2
C=O deformation	+39	+30
C—C—O bend	−45	−53
C—C stretch	−4	−4
CH$_2$2 rock	−48	−40
C—O stretch	+9	+5
CH$_2$ scissors	0	0
C=C stretch	−10	−12
C=O stretch	+12	+12
C—C torsion	−4	−10
C=O wag	+1	0
C-O torsion	−35	−41
C—H & C—O wag	+3	−7
CH$_2$ twist	+2	0
CH$_2$ wag	0	−6

[a] Ref. [16]. Frequency differences in cm^{-1}. Set 1 increased intensity on ultraviolet irradiation relative to Set 2.

butadiene. Table II shows a comparison of the observed differences between corresponding vibrational modes in Set 1 and Set 2 and the computed differences between the *cis*- and *trans*- conformers. The pattern of coincidence leaves no doubt of the assignment of Set 1 as arising from the *cis*- form of the molecule.

Solution Number 2: Accurate Calculation

The scaling procedure for prediction of vibrational frequencies works well as long as there are related molecules for which the spectral assignments are known so that the scaling factors can be determined and transferred to the unknown molecule. This covers a very wide range of organic molecules, but there are also exotic species, ions, inorganic molecules, and others for which closely related molecules cannot be identified with any degree of certainty. If accurate vibrational force fields and spectral frequencies are desired for these, there seems to be no alternative to the brute force method of simply doing the calculations at the required level of rigor. This is not an easy task.

Accurate absolute calculation of vibrational frequencies presents two major problems. First is the determination of the vibrational potential energy hypersurface to a sufficiently high degree of accuracy and over a wide enough range of displacement of the vibrational coordinates. Next is the problem of determining the eigenvalues of the system constrained to move on that surface (the anharmonic vibrational energy levels). These two parts of the overall problem may be discussed separately.

In principle, determination of the potential energy function for the vibrating molecule is straightforward. The electronic Schrödinger equation must be solved with the nuclei fixed at the equilibrium geometry around which the system is vibrating and then at displaced positions, carefully selected, to form a multidimensional grid of energy points as a function of nuclear coordinates. These computed energies may be substituted in the usual power series expansion of energy as a function of nuclear displacements

$$E = E_{\text{ref}} + \sum_i g_i q_i + \frac{1}{2} \sum_i \sum_j F_{ij} q_i q_j + \frac{1}{6} \sum_i \sum_j \sum_k F_{ijk} q_i q_j q_k \cdot + \cdots$$

where $q_i, q_j, q_k, \ldots$ represent vibrational coordinates, g_i is the force acting along coordinate i (all such forces being zero if the reference geometry is the equilibrium geometry), F_{ij} is a matrix of harmonic force constants, and F_{ijk} and similar higher order terms represent the anharmonic force and corresponding interaction constants. Termination of the series after the quadratric term gives the usual harmonic oscillator (Hooke's law) vibrational approximation. Unfortunately, this is a slowly converging series and to obtain vibrational frequencies accurate on the order of, say, 10 cm^{-1} requires the determination of at least some diagonal terms through sixth order, F_{iiiiii}. There are a tremendous number of terms at even the full fourth order if all coupling constants are included.

Aside from the great detail needed in the vibrational energy expansion, the problem is made more difficult by the high level at which the quantum chemical calculation must be carried out to achieve this level of accuracy. A large Hartree-Fock basis set is essential, probably triple zeta with two sets of d functions and preferably a set of f functions on the heavy atoms with a corresponding treatment of hydrogens. In addition a very good treatment of electron correlation is needed – Møller–Plesset perturbation theory through fourth order with use of the Davidson correction or the newer quadratic CI is probably barely adequate. Further complication can arise if the vibrational amplitude for the state being considered is sufficiently wide that it samples regions of the potential surface that are not adequately treated by the single determinant wavefunction presumably being used at the equilibrium position. All of this is intended to indicate that computation of the vibrational frequencies of molecules containing more than two or three atoms to an absolute accuracy of 10 cm^{-1} is at the very frontier of what is possible with present computational technology.

The computational problem has been somewhat eased by the recent development of analytical methods for determining higher derivatives, e.g. F_{ijk}. The group working with Professor N.C. Handy at Cambridge University has made especially valuable contributions in this area [17]. The techniques are still limited in the order of the derivative which can be taken and, more importantly, in the type of wavefunction with which they can be used. Rapid progress can be expected in this area, however. Even if only lower order derivatives can be taken, their use is helpful in limiting the number of displaced geometries at which calculations must be made, provided they can be used with a sufficiently high-level wavefunction.

While sufficiently accurate calculation of the energy surface is still the limiting factor in the accuracy with which absolute anharmonic vibrational frequencies can be calculated, it is worthwhile to look at the second part of the problem, the determination of the vibrational energy levels on that surface. This is no longer the trivial matter that it is within the harmonic oscillator approximation. Perturbation methods have been used in the past, especially for diatomic molecules and treatment of modes showing Fermi resonance. Lee *et al.* [17] have recently pursued this approach much further, applying it to the cyclopropenyl cation, $C_3H_3^+$. The technique is relative easy to use, but it has been applied only to low orders of the vibrational potential energy expansions and it cannot give accuracy as high as that being discussed here. When less accuracy is required, or when some specific feature such as interaction between a given pair of modes is being considered, the method is very promising.

A more difficult but more flexible approach is to seek a variational solution to the vibrational Schrödinger equation. The applicable techniques are very similar to the SCF and CI methods used in solving the electronic Schrödinger equation, so that much of the computational methodology developed for that area can be modified to treat the vibrational problem. Early work using this technique includes the study of Whitehead and Handy on H_2O [18], that of Botschwina on H_2O, HNO, HOF, and HOCl [19], of Murrell, Carter, and Halonen on HCN [20], and of Romanowski, Bowman, and Harding on HCHO [21]. In an effort to apply the technique to larger molecules, more recent studies have made use of the computer program anhar written by Péter Pulay [22].

The variational solution of the vibrational Schrödinger equation has been described in several publications [23–27], but it may be summarized as follows. First, the kinetic energy operator, most easily expressed in rectilinear coordinates, and the potential energy, commonly obtained in curvilinear coordinates, must be combined into a single Hamiltonian. In recent work [23–27], the potential energy expression is mapped onto rectilinear coordinates by a least-squares method. Considerable caution must be exercised not to lose accuracy in this step. Next, a vibrational CI method is used, in which the vibrational Hamiltonian is solved in a basis of harmonic oscillator product functions. Singly, doubly, triply, and quadruply substituted configurations are considered.

The heart of the method, and the most time-consuming part of the calculation, is a multi-step basis function selection procedure. First, the highest quantum number to be considered for each mode and each level of excitation (S, D, T, Q) is chosen on the basis of experience and trial. This gives the virtual space for the calculation, which was as large as 150,000 per run [26] for $C_3H_3^+$. Next a subspace of 10 to 20 of those configurations expected to contribute most strongly to the desired states is chosen. Diagonalization in this subspace permits identification of the states. The CI coefficients are estimated by first-order perturbation theory. This yields the primary space. Using the wavefunction in the primary space, the full basis is scanned by second-order perturbation theory to get the secondary space. This was still as large at 1588 per run in the study of C_3H_3+. The Davidson iterative method [28] was then used to get the desired eigenvalues. In a given run, from 3 to 10 energy levels could be obtained.

The method just described has been applied to HCN [23], to CH_3F and CD_2HF [24], to CD_3H [25], to the cyclopropenyl cation $C_3H_3^+$ and all of its deuterated derivatives [26], and to the perfluorocyclopropenyl cation [27], $C_3F_3^+$. The original references should be consulted for the detailed results. Fundamental and first overtone vibrational frequencies were derived, and in some cases, extensive families of overtone and combination bands were studied. Accuracy in every case appears to have been limited only by the accuracy of the vibrational potential surface available.

Several comments should be made about full anharmonic oscillator calculations. The harmonic oscillator approximation is a wonderfully useful although basically incorrect tool. It describes a simple world with normal modes of vibration, strict selection rules, hot bands coincident in frequency with fundamentals, overtones at exact multiples of the fundamentals, and no complications such as Fermi resonances. Unfortunately, that is not the world in which real molecules vibrate. A variational anharmonic calculation can include all real effects [29]. For example, Fermi resonances are reproduced automatically. In $C_3D_3^+$, as one illustration, a band is predicted to lie at 2569 cm^{-1} using the harmonic oscillator approximation. The full treatment shows that it is not only shifted in frequency by anharmonicity but also split into a doublet at 2537 and 2452 cm^{-1} by Fermi resonance with the overtone of a fundamental at 1257 cm^{-1}. The doublet is observed by a matrix isolation experiment at 2528 and 2432 cm^{-1} As another example, in $C_3HD_2^+$, the ν_3 fundamental is predicted to be split by 33 cm^{-1} and a splitting of 29 cm^{-1} is observed. Note that these Fermi resonances are found automatically with no special effort required.

A further advantage of an anharmonic vibrational calculation is that the identification of bands is more complete, including admixtures of overtone bands. As an illustration, CH_3F has three bands at 2863, 2926, and 2966 cm^{-1}. No one of these can be correctly characterized as a fundamental since each is an essentially equal mixture of ν_1, $2\nu_2$, and $2\nu_5$.

A heavy price is paid for all of the additional information that comes out of a variational treatment of vibrational anharmonicity. The computation for a species such as $C_3F_3^+$ strains the capacity, both in CPU time and in storage capacity, of any presently existing supercomputer. Of course, somewhat larger molecules can be treated if less stringent requirements are placed on accuracy. The work can be further simplified if all vibrational modes are not of interest and if highly excited states are not required.

Recommendations

Since the calculation of absolute anharmonic frequencies with an accuracy of 10 cm^{-1} or better for all vibrational modes is currently possible only for molecules containing no more than half a dozen atoms, this approach is not a choice for the vast majority of molecules of chemical interest. This is not to say that there are not numerous species worthy of such study, especially those small molecules and fragments that occur in high temperature environments where information about the anharmonic contributions to their thermodynamic properties is most urgently needed. For most molecules, however, the present state of the art permits only less accurate predictions of vibrational properties or, in very many cases, accurate prediction using scaling procedures based on the transferability of computational error and anharmonicity for a given type of molecular motion.

The present limit in size of molecules for which a detailed scaled force field approach at the *ab initio* level can be used is on the order of 30 atoms. This limit still excludes most known organic compounds, but it is sufficiently large to encompass type compounds that can be used to answer fundamental questions of molecular flexibility and response to mechanical deformation forces. It is also certainly as large or larger than the molecular size limit on compounds for which detailed vibrational assignments can be made from experiment.

Acknowledgements

The portion of the research described in this paper that was performed at The University of Texas has been supported by grants from The Robert A. Welch Foundation, Cray Research, Inc., and the Texas Advanced Technology Program. Related calculations have been performed on the Cray computers of the University of Texas Center for High Performance Computing.

References

[1] A. Somogi, G. Jalsovszky, C. Fülöp, J. Stark, and J. E. Boggs: *Spectrochim. Acta* **45A**, 679 (1989).
[2] Y. Xie, K. Fan, and J. E. Boggs: *Mol. Phys.* **58**, 401 (1986).
[3] P. Pulay, G. Fogarasi, F. Pang, and J. E. Boggs: *J. Am. Chem. Soc.* **101**, 2550 (1979).
[4] P. Pulay, G. Fogarasi, G. Pongor, J. E. Boggs, and A. Vargha: *J. Am. Chem. Soc.* **105**, 7037 (1983).
[5] P. Pulay, G. Fogarasi, and J. E. Boggs: *J. Chem. Phys.* **74**, 3999 (1981).
[6] G. Pongor, P. Pulay, G. Fogarasi, and J. E. Boggs: *J. Am. Chem. Soc.* **106**, 2765 (1984).
[7] G. Pongor, P. Pulay, G. Fogarasi, and J. E. Boggs: *J. Mol. Spectrosc.* **114**, 455 (1985).
[8] H. Sellers, P. Pulay, and J. E. Boggs: *J. Am. Chem. Soc.* **107**, 6487 (1985).
[9] Z. Niu, K. Dunn, and J. E. Boggs: *Mol. Phys.* **55**, 421 (1985).
[10] Y. Xie and J. E. Boggs: *J. Computat. Chem.* **7**, 158 (1986).
[11] G. Fogarasi and A. G. Császár: *Spectrochim. Acta* **44A**, 1067 (1988).
[12] A. G. Császár and G. Fogarasi: *Spectrochim. Acta* **45A**, 845 (1988).
[13] A. G. Császár and G. Fogarasi: *J. Phys. Chem.* **93**, 7644 (1980).
[14] R. F. Lemert and K. Dunn: *Thirteenth Austin Symposium on Molecular Structure*, Austin, Texas, March 12–14 1990.
[15] S. W. Charles, F. C. Cullen, N. L. Owen, and G. A Williams: *J. Mol. Struct.* **157**, 17 (1987).
[16] K. Fan and J. E. Boggs: *J. Mol. Struct.* **157**, 31 (1987).
[17] T. J. Lee, A. Willetts, J. F. Gaw, and N. C. Handy: *J. Chem. Phys.* **90**, 4330 (1989).
[18] R. J. Whitehead and N. C. Handy: *J. Mol. Spectrosc.* **59**, 459 (1976).
[19] P. Botschwina: *Chem. Phys.* **40**, 33 (1979).
[20] J. N. Murrell, S. Carter, and L. O. Halonen: *J. Mol. Spectrosc.* **94**, 307 (1982).
[21] H. Romanowski, J. M. Bowman, and L. B. Harding: *J. Chem. Phys.* **82**, 4155 (1985).
[22] Program ANHAR, has been combined with other vibrational programs by Kevin Dunn into a shell script suitable for use on supercomputers with a UNIX operating system and is available from the present author.
[23] K. M. Dunn, J. E. Boggs, and P. Pulay: *J. Chem. Phys.* **85**, 5838 (1986).
[24] K. M. Dunn, J. E. Boggs, and P. Pulay: *J. Chem. Phys.* **86**, 5088 (1987).
[25] K. M. Dunn: *Chem. Phys. Lett.* **139**, 165 (1987).
[26] Y. Xie and J. E. Boggs: *J. Chem. Phys.* **90**, 4320 (1989).
[27] Y. Xie and J. E. Boggs: *J. Chem. Phys.* **91**, 1066 (1989).
[28] E. R. Davidson: *J. Comp. Phys.* **17**, 85 (1975).
[29] The studies described in Refs. [23–27] did not include rotational coupling, but this is easily added.

THEORETICAL APPROACH TO TORSIONAL BAND STRUCTURES OF SOME NON-RIGID MOLECULES

YVES G. SMEYERS
Instituto de Estructura de la Materia (CSIC)
c/ Serrano 123
28006 Madrid
Spain

and

DAVID C. MOULE
Chemistry Department
Brock University
St. Catharines
Ontario
L2S 3A1
Canada

ABSTRACT. Ab initio theoretical calculations for determining the torsional band structure in the electronic spectrum, as well as in the far infrared spectrum are presented. In a first example, the band structure due to the methyl torsion and aldehydic hydrogen wagging modes in the $\tilde{a}^3 A''(n\pi^*) \leftarrow \tilde{X}^1 A'(n^2)$ electronic spectrum is considered. In a second example, the band structure due to the double methyl rotation in the far infrared spectrum of acetone-h_6 is deduced. In both examples, Group Theory for Non-rigid Molecules is widely employed not only for simplifying the solution of the nuclear motion equations, but also for classifying the energy levels and assigning the transitions. A very good agreement with experiment is found in both cases.

1. Introduction

A non-rigid molecule is a molecular system which presents large amplitude vibration modes. This kind of motion appears whenever the molecule possesses various isomeric forms of similar energy, separated by relatively low energy barriers. In such cases, molecular transformation movements from one isomeric form to another occur.

This molecular deformability confers interesting properties to this kind of molecules: such as a large specific heat capacity, etc.. In addition, their far Infrared spectra present a prolific band structure. Furthermore, small electronic excitations produce large structural changes in the excited states. As a result, because of the Franck-Condon factors, long band progressions are observed in their electronic spectra.

All these types of phenomena can be studied theoretically. For this purpose, the potential energy hypersurfaces on which the nuclei are moving have to be determined, the

L. A. Montero and Y. G. Smeyers (eds.), Trends in Applied Theoretical Chemistry, 199–211, 1992.
© 1992 *Kluwer Academic Publishers.*

Schrödinger equations for the nuclear motions solved, and the level populations, band locations and band intensities deduced from the eigenvalues and eigenvectors.

In the following, we give two examples of inverse spectroscopy calculations. By inverse spectroscopy, we mean a technique which simulates completely the spectrum of a molecule from ab initio calculations.

In particular, we present the theoretical determination of band structure of the absorption spectrum $\tilde{a}^3 A''(n\pi^*) \leftarrow \tilde{X}^1 A'(n^2)$ of thioacetaldehyde. In the same way, we present the far infrared spectrum of acetone.

In all these calculations, Group Theory for Non-Molecules may be conveniently used not only for simplifying the calculations, but also for classifying and labelling the transitions.

2. Theory

2.1. NON-RIGID GROUPS FOR THIOACETALDEHYDE AND ACETONE

Each nuclear transformation movement on the potential energy surface may be described by some operator. On the other hand, all the nucleus movements can be accounted by some nuclear Hamiltonian operator. As is well known, the complete set of the transformation operators, which commute with such an Hamiltonian operator, form a *group*, we call the *Non-Rigid Group* [1].

Such transformation movements can be generally separated from the external rotation, at least in a first approach, whenever the transformation can be described in terms of rotations around some axes supported by a solid frame. In this special case, these movements may be conveniently described by physical operations such as internal rotations [1, 2].

When the reference frame is reduced to a single atom, the separation in internal and external motions is not possible. The transformation movements are then more easily described by permutations and permutation-inversions of identical nuclei, such as in the Longuet-Higgins formalism [3].

Thioacetaldehyde in its lowest excited electronic states presents essentially two degrees of freedom of large amplitude: the torsion of the methyl group and the wagging of the aldehydic hydrogen atom. Both motions may be described as a rotation, of orders three and one for the torsion and wagging, respectively. In this paper, the other degrees of freedom are assumed to be of small amplitude and thus separable.

The Hamiltonian operator for such a system can then be written as:

$$H = - \left[\frac{\partial}{\partial\theta} B_{11} \frac{\partial}{\partial\theta} + \frac{\partial}{\partial\theta} B_{12} \frac{\partial}{\partial\alpha} + \frac{\partial}{\partial\alpha} B_{21} \frac{\partial}{\partial\theta} + \frac{\partial}{\partial\alpha} B_{22} \frac{\partial}{\partial\alpha} \right] + V(\theta, \alpha) \tag{1}$$

where θ and α are the coordinates for the torsion and wagging, respectively, B_{ij} the so called rotation constants can be calculated from the geometry of the molecule in the electronic state considered [4].

The Hamiltonian operator (1) is seen to be invariant under:
1) A rotation of order three of the methyl group;
2) A simultaneous change of the sense of both rotations.

The correspondant subgroup are:

$$C_3^I = [\hat{E} + \hat{C}_3 + \hat{C}_3^2] \tag{2}$$

and

$$V^I = [\hat{E} + \hat{V}] \tag{3}$$

where $\hat{C}_3$ and $\hat{C}_3^2$ are rotations of order three of the methyl group:

$$\hat{C}_3^n f(\theta, \alpha) \equiv f(\theta + 2n\pi/3, \alpha) \tag{4}$$

and $\hat{V}$ the inversion or double switch operator:

$$\hat{V} f(\theta, \alpha) \equiv f(-\theta, -\alpha) \tag{5}$$

As a result, the non-rigid group for the torsion of the methyl group and wagging of the aldehydic hydrogen atom in thioacetaldehyde can be written as:

$$G_6 = C_3^I \wedge V^I \sim C_{3v} \tag{6}$$

which is a group of order six isomorphic to the symmetry point group C_{3v}.

Acetone presents essentially in its singlet ground state two degrees of freedom of large amplitude: the torsion of both methyl groups. Notice that the wagging of the oxygen atom is only important in the excited states, therefore it may be disregarded in the ground state.

The Hamiltonian operator for the double internal rotation in acetone can be written as:

$$H = -\left[\frac{\partial}{\partial \theta_1} B_{11} \frac{\partial}{\partial \theta_1} + 2 \frac{\partial}{\partial \theta_1} B_{12} \frac{\partial}{\partial \theta_2} + \frac{\partial}{\partial \theta_2} B_{22} \frac{\partial}{\partial \theta_2} \right] + V(\theta_1, \theta_2) \tag{7}$$

where θ_1 and θ_2 are the two torsional angles. B_{ij} are the rotational constants that can be calculated from the geometrical data [5].

This operator is seen to be invariant under:

1) A rotation of order three of each of the methyl groups;
2) A simultaneous change of the sense of the rotations of both rotors;
3) An exchange of the rotation coordinates in which the two methyl groups are indiscernible.

The corresponding subgroups are C_3^I for each methyl group (2), the double switch subgroup, V^I (3), and the exchange subgroup, W^I, which is defined as:

$$W^I = [\hat{E} + \hat{W}] \tag{8}$$

In this expression, $\hat{W}$ is the exchange operator:

$$\hat{W} f(\theta_1, \theta_2) \equiv f(\theta_2, \theta_1) \tag{9}$$

As a result, the non-rigid group for the double rotation in acetone is written as:

$$G_{36} = (C_3^I \times C_{3'}^I) \wedge (V^I \times W^I) \tag{10}$$

which is a group of order 36, that has no analogue with the rigid symmetry point groups [5].

TABLE I
Character Table for the internal rotation and wagging in thioacetaldehyde

	$\hat{E}$	$2\,\hat{C}_3$	$3\,\hat{V}$
A_1	1	1	1
A_2	1	1	-1
E	2	-1	0

2.2. THE SYMMETRY EIGENVECTORS FOR THE G_6 AND G_{36} NON-RIGID GROUPS

In order to solve the Schrödinger equation for the torsion and wagging in thioacetaldehyde and for the double torsion in acetone, the torsional solutions are developed in terms of the eigenfunctions of the double free rotor equations. The coefficients of such an expansion are determined by solving the secular equation, i.e., by diagonalizing the Hamiltonian matrix, the elements of which are calculated between the basis functions.

When the basis functions are the symmetry eigenvectors, which transform according to one of the irreducible representations of the Hamiltonian symmetry group, the Hamiltonian matrix is factorized into boxes. This feature may be advantageously used for simplifying the solution, as well as for classifying the torsional wavefunctions. For this reason, the symmetry eigenvectors will be employed in the following.

The symmetry eigenvectors for the G_6 non-rigid group of thioacetaldehyde are easily deduced from the character table (Table I) of this group. This table is identical to that of the C_{3v} point group, except for the operations:

From this character table, the following symmetry eigenvectors are easily deduced on the basis of the double free rotor solutions:

$$\chi_{A_1} = \cos 3k\theta \cos l\alpha; \qquad \sin 3k\theta \sin l\alpha$$

$$\chi_{A_2} = \cos 3k\theta \sin l\alpha; \qquad \sin 3k\theta \cos l\alpha$$

$$\chi_E = \begin{cases} \cos(3k \pm 1)\theta \cos l\alpha; & \sin(3k \pm 1)\theta \sin l\alpha \\ \cos(3k \pm 1)\theta \sin l\alpha; & \sin(3k \pm 1)\theta \cos l\alpha \end{cases} \tag{11}$$

The character table for the G_{36} group of acetone is more involved, and can be found in [1, 2, 3]. We reproduce it in Table II.

As it can be seen, the G_{36} group contains nine irreducible representations, four non-degenerate, four doubly degenerate, and one fourfold degenerate representations. From this character table, the symmetry eigenvectors cannot be deduced easily as before by simple intuition, they have to be calculated by using the well known projection operators:

$$\hat{P}_j = \frac{l_j}{N} \sum_R^N \chi_j(\hat{R}) \tag{12}$$

TABLE II

Character table for the double internal rotation in planar acetone with interaction between the rotors

	E	$2C_3$ $2C_{3'}$	2 $C_3C_{3'}$	2 $C_3C_{3'}^2$	W $2W\times$ $C_3C_{3'}^2$	$6W\times$ C_3C_{13}	WV $2WV\times$ $C_3C_{3'}$	$6WV\times$ $C_3C_{3'}$	V $8V\times$ $C_3C_{3'}$
A_1	1	1	1	1	1	1	1	1	1
A_2	1	1	1	1	-1	-1	1	1	-1
A_3	1	1	1	1	1	1	-1	-1	-1
A_4	1	1	1	1	-1	-1	-1	-1	1
E_1	2	-1	2	-1	0	0	2	-1	0
E_2	2	-1	2	-1	0	0	-2	1	0
E_3	2	-1	-1	2	2	-1	0	0	0
E_4	2	-1	-1	2	-2	1	0	0	0
G	4	1	-2	-2	0	0	0	0	0

In this expression, N is the order of the group, l_j the dimension of the irreducible representation j on which the operator projects, χ_j is the character of the representation for the operation R, being $\hat{R}$ the corresponding operator. The sum runs over all the operations of the group. These symmetry eigenvectors are given in references [1,2] and [5].

3. Potential Energy Calculations

3.1. THIOACETALDEHYDE

Since HF calculations for both ground and first excited states of acetaldehyde seemed to yield reasonable results, the same level of approximation was employed in the present calculations [6]. The potential energy function for thioacetaldehyde was determined by using the RHF and UHF schemes for the singlet ground state, and first triplet excited states, respectively.

The basis set employed through all the calculations was 4-31G, complemented by six d-type gaussian orbitals with exponent, a=0.39, centered on the sulphur atom.

The geometrical parameters reported for the thioacetaldehyde ground state [7] were used in the conformational calculations of this state. In exchange, the geometry of the triplet excited state was partially optimized, assuming that the transition $(n \rightarrow \pi^*)$ affected only the C=S bond length, and the C—C—S bond angle. The following parameters were found r(C=S)=1.80 Å, and α(CCS)=122°, in relatively good agreement with theoretical values found for thioformaldehyde [8,9]. The C=S theoretical bond length, however, is seen to be too long when compared with the experimental value [10].

Using these geometries, calculations were performed for both singlet and triplet states, in which the methyl group was rotated about the C—C bond by an angle of 30° from the eclipsed conformation. Simultameously the wagging angle of the aldehydic hydrogen atom was modified from –35° to +35° for the singlet state, and from –60° to +60° for the triplet state.

The results of each of these calculations were plotted in an A_1 symmetry adapted double

Fourier expansion (11), in terms of the rotation angle (θ) and wagging angle (α).

The following expression was obtained (in cm^{-1}) for the singlet ground state:

$$V(\theta, \alpha) = 20,703.8 - 20,486.8 \cos \alpha + 76.3 \cos 3\theta \cos \alpha -$$

$$-302.2 \ \cos 3\theta \cos 2\alpha + 657.1 \sin 3\theta \sin \alpha \qquad (13)$$

and for the first triplet excited state ($n\pi^*$):

$$V(\theta, \alpha) = 9,803.6 - 1,101.8 \cos \alpha + 1,356.5 \cos 3\alpha +$$

$$+524.2 \cos 3\theta \cos \alpha - 492.7 \cos 3\theta \cos 2\alpha +$$

$$+30.1 \ \cos 3\theta \cos 3\alpha - -10.7 \sin 3\theta \sin \alpha + 159.9 \sin 3\theta \sin 3\alpha. \qquad (14)$$

The presence of large mixed terms in this last expansion suggests the existence of large coupling between both vibrational modes in the triplet state.

An analysis of these potential energy functions shows that thioacetaldehyde presents a planar structure and an eclipsed conformation in the singlet ground state. On the contrary, thioacetaldehyde exhibits a pyramidal structure and an antieclipsed conformation in the first triplet excited state ($n \rightarrow \pi^*$). This important conformational and structural change suggests that the electronic spectrum of thioacetaldehyde must exhibit large band structure because of the Franck–Condon principle.

3.2. ACETONE

There have been several theoretical calculations for determining the potential energy function for the double rotation in acetone [11,12]. The first detailled analysis of the torsional of the torsional dynamics was made by Smeyers et al. in 1981 [5]. In this study, the total energy was mapped out by using the CNDO/2 semi-empirical method as a function of the two rotational angles. From the potential energy function, they calculated the ν_{17} and ν_{12} modes to be 91.99 and 67.36 cm^{-1}, respectively, which indicates that the derived potential was too shallow. In a later study, Smeyers and Huertas, revised the barrier and saddle points upwards to a more realistic values using ab initio calculations [13].

Similar calculations have been reported. Crighton and Bell [14] were able to calculate values for the two torsional fundamental levels in reasonable agreement with the observed infrared frequencies. They used a semi-rigid model for acetone in which the frame and the methyl coordinates were fixed at the optimized eclipsed-eclipsed equilibrium geometry.

Recently, Ozkabak, Philis and Goodman [15], using the torsional Hamiltonian of Crighton et al. evaluated the potential energy surface of acetone with a variety of different ab initio approaches, which included full optimization of the geometry, as well as the electronic correlation to some extent. These authors concluded that a signicant flexing of the molecular structure occurs when the methyl groups are rotating as a result of the steric hindrance.

In the present study, fully relaxed calculations (except for the oxygen atom which was confined in the CCC plane) were performed in the Restricted Hartree-Fock approximation complemented by a full Möller-Plesset perturbation theory calculation up to second order in order to introduced some electronic correlation effects. In these calculations, the 6–31 G (p,d) basis set was employed.

Seven data points conviently chosen were fitted to an A_1 symmetry adapted Fourier expansion in seven terms, which define the potential energy surface [16]. Next, we give the potential energy function for the double rotation in acetone obtained at the RHF + MP2 level, in cm^{-1}:

$$V(\theta_1, \theta_2) = 332.2345 - 200.757(\cos 3\theta_1 + \cos 3\theta_2) + 64.8700 \cos 3\theta_1 \, \cos 3\theta_2 -$$

$$-0.1625(\cos 6\theta_1 + \cos 6\theta_2) +$$

$$+2.3525(\cos 6\theta_1 \, \cos 3\theta_2 + \cos 3\theta_1 \, \cos 6\theta_2) - 0.0775 \cos 6\theta_1 \, \cos 6\theta_2 -$$

$$- \sin 3\theta_1 \, \sin 3\theta_2 \tag{15}$$

where the sine-sine term describes the gearing effect [5].

4. Schrödinger Equation Solution

4.1. THIOACETALDEHYDE

At this point, we have to solve separately two Schrödinger equations, corresponding to the Hamiltonian operator (1), for both the singlet ground and the first triplet excited states.

For this purpose the potential energy functions determined, in the RHF and UHF approximations for the singlet and triplet states, as well as the kinetic B_i constants, were introduced in expression (1) for the Hamiltonian operator. These kinetic constants were calculated into the rigid symmetric rotor approximation taking into account the structural changes between both states.

The kinetic B_i constants for each rotor, as well as the coupling terms between them, B_{ij}, were determined according to the following expressions, [5] and [17]:

$$B_1 = \frac{\hbar^2}{2} \frac{I_2}{I_1 I_2 - \Lambda_{12}^2}, \qquad B_{12} = \frac{\hbar^2}{2} \frac{\Lambda_{12}}{I_1 I_2 - \Lambda_{12}^2} \tag{16}$$

where I_i is the reduced moment of inertia of one of the rotor, and Λ_{ij} the coupling term defined as follows:

$$I_1 = A_1(1 - \sum_{i=a,b,c} A_1 \lambda_{1i}^2 / M_i), \qquad \Lambda_{12} = \sum_{i=a,b,c} A_1 A_2 \lambda_{1i} \lambda_{2i} / M_i \tag{17}$$

In these expressions, A_1 is the moment of inertia of the rotor 1, M_i the three moments of inertia of the whole molecule, and λ_{1i} and λ_{2i} the direction cosines between the internal rotation axes and the ith principal axis.

The torsional solutions were developed in terms of the symmetry eigenfunctions (11), equivalent to products of 31 trigonometric functions of the torsional angle, and 43 trigonometric functions of the wagging angle, giving rise to a Hamiltonian matrix of order 31×43 = 1333. Using the symmetry eigenfunctions (11), this Hamiltonian matrix is factorized into four boxes of dimensions 237, 236, 430 and 430, for the A_1, A_2, and doubly degenerate E representations, respectively.

4.2. ACETONE

For determining the torsional structure of acetone spectrum, we have to solve only the Schrödinger equation corresponding to the Hamiltonian operator (7) for the singlet ground state.

For this purpose, the potential energy functions determined in the RHF and RHF + MP2 calculations, as well as the kinetic B_{ij} parameters, were introduced in expression (7) for the Hamiltonian operator. In the present case, the kinetic B_{ij} constants were allowed to vary with the rotation, since the molecular geometry was fully optimized. Therefore, the values obtained for each conformation by using equations (16) and (17) were fitted to a A_1 symmetry adapted functional form identical to that of the potential energy function (15).

The torsional solutions for the double rotation in acetone were developed in terms of the symmetry eigenvectors given in reference (1), equivalent to products between 31 trigonometric functions of the torsional angle of each of the rotors, giving rise to a Hamiltonian matrix of order $31 \times 31 = 961$. By the use of the symmetry eigenvectors, this Hamiltonaian matrix is factorized into sixteen boxes of dimensions 36, 30, 30, 25, for the non-degerate representations A_1, A_2, A_3, and A_4, 55, 45, 55, 45, for the doubly degenerate representations E_1, E_2, E_3, and E_4, and 110 for the fourfold degerate representation G.

5. Spectrum Calculations

5.1. TORSIONAL BAND STRUCTURE DETERMINATION OF ELECTRONIC SPECTRA

The solution of the Schrödinger equation for the nuclear motion furnishes the vibrational levels and wave-functions corresponding to a given electronic state. If we further consider electronic excited states, band structures of electronic spectra may be determined. As is well known, the square of the overlap between the vibrational functions of both states will give us the so-called *Franck-Condon* factors which are a measure of the intensity of the transition between the vibrational levels of both states:

$$I_{if} \approx C_i \, g \, \langle \Psi_i \mid \Psi_f \rangle^2 \tag{18}$$

where C_i is the population, Ψ_i and Ψ_f the torsional wave-functions, the indexes i and f stand for the initial and final states, and g is the nuclear statistical weight.

The Franck-Condon factor determination is of special interest when the two electronic states, involved in the transition exhibit very different geometries. This is especially true in the case of electronic transitions in the valence shell such as $n \rightarrow \pi^*$, which induces conjugation change, as well as geometrical change, in the molecular system. This phenomenon was studied in the fluorescence spectra of acetaldehyde and acetone [18,19], and in the phosphorescence spectra of thioacrolein and thioacetaldehyde [20,21] and thioacetone [22].

5.2. TORSIONAL BAND STRUCTURE OF THE $\tilde{a}^3 A'' \leftarrow \tilde{X}^1 A'$ OF THIOACETALDEHYDE

In the following we shall deduced theoretically the $\tilde{a}^3 A'' \leftarrow \tilde{X}^1 A'$ torsional band structure of thioacetaldehyde as an example. Thioacetaldehyde exhibits a planar *trans* preferred

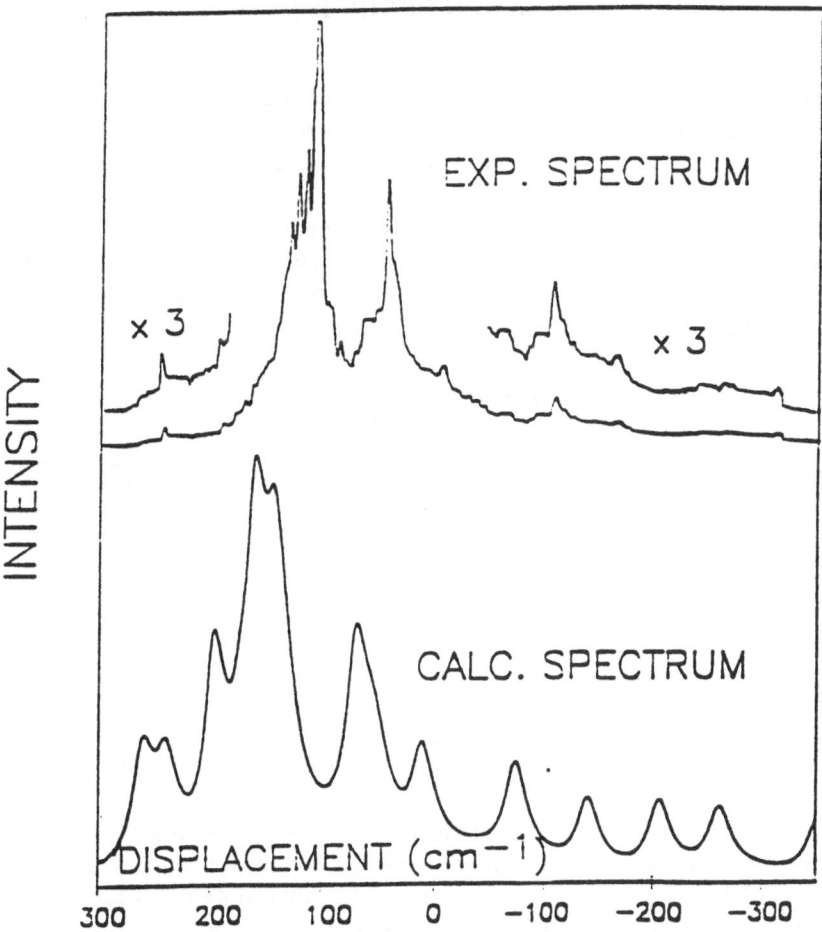

Fig. 1. Band structures corresponding to the methyl torsion and aldehydic hydrogen wagging modes in the $\tilde{a}^3 A'' \leftarrow \tilde{X}^1 A'$ spectrum of thioacetaldehyde. The broadness of the bands of the theoretical spectrum is simulated by a Lorentzian profile [23].

conformation in the singlet ground state, whereas it presents a non-planar *gauche* ones in the first triplet excited state ($n \rightarrow \pi^*$). The principal deformations involve the torsion of the methyl group and the wagging of the aldehydic hydrogen atom.

The differences between the eigenvalues of both equations will furnish the relative locations of the torsional bands, whereas the Franck-Condon factors (18) will give the relative intensities. Notice, that the nuclear statistical weights have to be taken into account to compare the intensities of transitions belonging to different degrees of degeneracy [23].

The band structures corresponding to the methyl torsion and hydrogen wagging modes in the $\tilde{a}^3 A'' \leftarrow \tilde{X}^1 A'$ spectrum of thioacetaldehyde is given in Figure 1, together with the experimental spectrum. A good agreement is observed [23]. The spectra for the partially and fully deutered species are given in the same reference.

5.3. THEORETICAL DETERMINATION OF THE TORSIONAL FAR INFRARED SPECTRUM OF ACETONE

Internal rotation in two threefold symmetric top molecules has been mainly studied by microwave spectroscopy [24, 25]. Some studies, however, exist in the Far Infrared spectroscopy (FIR) [26, 27]. Next, the band structure calculations of FIR of acetone will be developed as an example [28, 29].

The solution of the Schrödinger equation for the nuclear motions will provide energy levels and torsional wave-functions. Band locations will be given by energy differences and intensities by the square of the electric dipole moment variation with the transition, weighed by the difference of populations.

$$I \sim (C_i - C_f) \, g \, \langle \Psi_i | \, \mu(\theta_1, \theta_2) \, | \Psi_f \rangle^2 \tag{19}$$

where C_i and C_f are the populations, Ψ_i and Ψ_f the torsional wave-functions, the indexes i and f stand for the initial and final states, and g is the nuclear statistical weight.

5.4. SELECTION RULES

Group Theory for Non-Rigid Molecule (NRG), permits the classification of the torsional wave-functions according to the irreducible representations of the symmetry group of the molecule. As it is well known, the scalar product of (19) does not vanish when the direct products of the irreducible representations, under which Ψ_i and Ψ_f transform, contain at least one of the components of the dipole moment variation. Thus, when symmetry properties of these components are known, *Selection Rules* may be established.

In single rotor molecules such as toluene, phenol or F-phenols [30–31] where the NRG's coincide with some symmetry point groups, the deduction of the selection rules are straightforward. In double rotor molecules, this deduction is not so easy [28, 29].

As seen before (10) , the NRG of acetone is the G_{36}. The first step should be to determine the representations of the G_{36} group under which the electric moment components transform. For that purpose, let us consider acetone in a molecular axis system, in which the z axis coincides with the C_2 axis of the molecule, the y axis lies perpendicular to the z axis in the CCC plane, and the x axis stands perpendicular to the CCC plane.

With this definition, it is seen that the angle exchange operation $\hat{W}$ is equivalent to a C_2 rotation of the torsional angles, the $\hat{V}$ operation is equivalent to a reflexion in the zy plane, and the $\hat{W}\hat{V}$ operation to a reflexion in the zx plane. Let us now superimpose the non-degenerate representations of the G_{36} group, with their characters, on the representations of the symmetry point group C_{2v}, with their characters. This operations is done in Table III.

In the same table, the representations of the G_{36} and C_{2v} groups, in trigonometric basis sets are given, as well as the electric dipole moment components. Thus, it is seen that the z, x, and x electric dipole components transform as the A_1, A_2 and A_4 representations of the G_{36} group, respectively.

From the G_{36} NRG character table, Table II, and Table III, Selection Rules for the electric dipole moment transitions in the Far Infrared Spectrum of acetone may be easily deduced. They are given in Table IV.

TABLE III

Superimposition of the acetone G_{36} NRG (capital letters) and the C_{2v} symmetry point (small letters) character tables, with their representations in trigonometric basis sets, as well as the electric dipole moment components

		E	C_2	ZX	ZY	C_{2v}	G_{36}	μ
		E	W	WV	V	repre.	representations	
a_1	A_1	1	1	1	1	$cos 2\theta$	$\cos 3\theta_1 + \cos 3\theta_2$	T_z
a_2	A_3	1	1	-1	-1	$\sin 2\theta$	$\sin 3\theta_1 + \sin 3\theta_2$	R_z
b_1	A_2	1	-1	1	-1	$\cos \theta$	$\sin 3\theta_1 - \sin 3\theta_2$	T_x
b_2	A_4	1	-1	-1	1	$\sin \theta$	$\cos 3\theta_1 - \cos 3\theta_2$	T_y

TABLE IV

Selection Rules for the electric dipole transitions in the FIR spectrum of acetone

$\mu_z(A_1)$	$\mu_y(A_4)$	$\mu_x(A_2)$
$A_1 \leftrightarrow A_1$	$A_1 \leftrightarrow A_4$	$A_1 \leftrightarrow A_2$
$A_2 \leftrightarrow A_2$	$A_2 \leftrightarrow A_3$	$A_3 \leftrightarrow A_4$
$A_3 \leftrightarrow A_3$		
$A_4 \leftrightarrow A_4$		
$E_1 \leftrightarrow E_1$	$E_1 \leftrightarrow E_1$	$E_1 \leftrightarrow E_1$
$E_2 \leftrightarrow E_2$	$E_3 \leftrightarrow E_4$	$E_2 \leftrightarrow E_2$
$E_3 \leftrightarrow E_3$		$E_3 \leftrightarrow E_4$
$E_4 \leftrightarrow E_4$		
$G \leftrightarrow G$	$G \leftrightarrow G$	$G \leftrightarrow G$

5.5. FAR INFRARED SPECTRUM OF ACETONE

In Table IV, many transitions appear to be allowed. We have to consider, however, that acetone may be regarded, at least in a first approach, as an oblate symmetrical top, with the largest moment, I_c, parallel to the x axis. In such circumstances, the overall rotation selection rules will be $\triangle J = 0, \pm 1$ and $\triangle K = 0$, for the transitions along the x axis, and $\triangle J = 0, \pm 1$ and $\triangle K = \pm 1$, for the transitions in the zy plane, in the symmetric top approximation. As a result, the transitions along the x axis will give rise the a sharp band of C type, while the transitions in the zy plane will give rise to broad flat bands. Up to now, only the sharp bands of C type can be observed. Thus, in the following, only the transitions along the x axis will be considered, i.e., those of the column on the right of Table IV.

The torsional far infrared spectrum of acetone has been recently studied by using the symmetry point groups [15]. Intensities, however, were not calculated.

The FIR spectrum of acetone, in gas phase, has been measured by several authors [32,33]. Two clusters of bands have been observed about 124 cm^{-1} and 102 cm^{-1}. The first cluster is formed by three bands at 125.16, 124,61 and 122.04 cm^{-1}, with the central band the strongest. The second cluster shows a somewhat more involved structure. Three bands can be distinguished at 102.92, 104.02 and 105.44 cm^{-1}. Here, again the central band is the strongest.

The first cluster can be associated to the methyl rotation clockwise – counterclockwise ν_{17} mode, $v = 0 \rightarrow v = 1$. This cluster is calculated theoretically to result from the transitions: $A_1 \rightarrow A_2$, at 130.67 cm^{-1} (0.472), $G \rightarrow G$, at 129.45 cm^{-1} (0.908), $E_1 \rightarrow E_1$, at 128.49 cm^{-1} (0.173), and $E_3 \rightarrow E_3$, at 128.47 cm^{-1} (0.142). The intensities given between parentheses have to be multipied by 10^{-4}.

The second cluster can be associated to the first sequence, $v = 1 \rightarrow v = 2$. This is calculated to result from the transitions: $E_1 \rightarrow E_1$, at 113.70 cm^{-1} (0.068), $E_4 \rightarrow E_3$, at 112.49 cm^{-1} (0.060), $G \rightarrow G$, at 111.87 cm^{-1} (0.323), and $A_2 \rightarrow A_1$, at 110.32 cm^{-1} (0.222).

The calculated intensity distribution appears to reproduce the general observed patterns of bands among these fundamental and first sequence series. In the acetone spectrum two components are resolved with the predicted intensities, the third component is accounted for by the superimposition of the $E_n \rightarrow E_m$ transitions.

6. Conclusions

In this paper, we gave two examples of theoretical determinations of torsional band structures. We have successfully determined from ab initio calculations the band structure of the absorption spectrum $\tilde{a}^3 A''(n\pi^*) \leftarrow \tilde{X}^1 A'(n^2)$ of thioacetaldehyde. In the same way , we have deduced the far infrared spectrum of acetone, which is compared favorably with the experiment.

In all these calculations, Group Theory for Non-Rigid Molecules was conveniently used not only for simplifying the calculations, but also for classifying and labelling the transitions.

The techniques described in this paper represent a step in the calculations of the inverse spectra of medium size molecules, which possess large amplitude vibrational mode. With increases in the computer speed and computer memory, which is accompanied by a reduction of computing cost, the generation of the inverse spectra will become relatively straigthforward.

At the present time those calculations are still expensive in both computer time and manpower. Most of computer cost arises from the obtention of the potential energy surface. The prospect of massively paralle processors should ameliorate this aspect. It is expected that the need to generate computer software, which is especially designed for each molecule, will be replaced in a next future by a general all-purpose program.

References

[1] Y.G. Smeyers: *Teoría de Grupos para Moléculas No-rígidas. El Grupo Local. Aplicaciones*, in Memorias, Vol. 13, Real Academia de Ciencias, Madrid, 1989.
[2] Y.G. Smeyers: *Introduction to Group Theory for Non-rigid Molecules*, Adv. Quantum Chem., (P.O. Löwdin ed.), Academic P., New York, in press.
[3] H.C. Longuet-Higgins: *Mol. Phys.* **6**, 445 (1963).
[4] Y.G. Smeyers, M.N. Bellido, and A. Niño: *J. Mol. Struct. (Theochem)* **166**, 1 (1988).
[5] Y.G. Smeyers and M.N. Bellido: *Int. J. Quantum Chem.* **19**, 553 (1981).
[6] Y.G. Smeyers, A. Niño, and M.N. Bellido: *Theoret. Chim. Acta* **69**, 259 (1988).
[7] S. Paone, D.C. Moule, A.E. Bruno, and R.P. Steer: *J. Mol. Spectrosc.* **107**, 1 (1984).
[8] R.H. Judge, D.C. Moule, A.E. Bruno, and R.P. Steer: *J. Chem.Phys.* **87**, 60 (1987).
[9] M. Noble, E.C. Appel, and E.K.C. Lee: *J.Chem.Phys.* **78**, 2219 (1983).
[10] D.J. Cloutier and D.A. Ramsay: *Ann. Rev.Phys.Chem.* **34**, 31 (1983).

[11] M.H. Whango and S. Wolfe: *Can. J. Chem.* **55**, 2778, (1977);
    P. Bowers and L. Schafer: *J.Mol. Struct.* **69**, 233 (1980);
    L. Random, J. Baker, P.M. Gill, R.H. Nobes, and N.V. Riggs: *J. Mol. Struct.* **126**, 271 (1985).
[12] Y.G. Smeyers and M.N. Bellido: *Int. J. Quantum Chem.* **23**, 507 (1983).
[13] Y.G. Smeyers, A. Huertas-Cabrera, and S. Suhai: *Theoret. Chim. Acta* **64**, 97 (1983).
[14] J.S. Grighton and S. Bell: *J. Mol. Spectrosc.* **118**, 383 (1986).
[15] A.G. Ozkabak, J.G. Philis, and L. Goodman: *J. Am. Chem. Soc.* **112**, 7854 (1990).
[16] Y.G. Smeyers: *J. Mol. Struct. (Theochem)* **107**, 3 (1984).
[17] Y.G. Smeyers and A. Henandez-Laguna: in *Structure and Dynamics of Molecular Systems*, Daudel et al. (eds.), D. Reidel, Dordrecht, 1985, pp. 23–40.
[18] M. Noble and E.K.C. Lee: *J. Chem. Phys.* **81**, 1632 (1984).
[19] M. Baba, I. Hanazaki, and V. Nagashima: *J. Chem. Phys.* **82**, 3938 (1985).
[20] D.C. Moule, R.H. Judge, H.L. Gordon, and J.D. Goddard: *Chem. Phys.* **105**, 97 (1986).
[21] R.H. Judge, D.C. Moule, A.E. Bruno, and R.P. Steer: *J. Chem. Phys.* **87**, 60 (1987).
[22] D.C. Moule, Y.G. Smeyers, M.L. Senent, D.J. Clouthier, J. Karolczak, and R.H. Judge: *J. Chem. Phys.* **95**, 3137 (1991).
[23] Y.G. Smeyers, A. Niño, and D.C. Moule: *J. Chem. Phys.* **93**, 5786 (1990).
[24] J.D. Swalen and C.C. Costain: *J. Chem. Phys.* **31**, 1561 (1959).
[25] P. Groner and J.R. Durig: *J. Chem. Phys.* **66**, 1856 (1977).
[26] J. Meier, A. Bauder, and Hs.H. Günthard: *J. Chem. Phys.* **57**, 1219 (1972).
[27] J.R. Durig, Y.S. Li, and P. Groner: *J. Mol. Spectrosc.* **62**, 159 (1976).
[28] Y.G. Smeyers, A. Hernandez-Laguna, M.N. Bellido, and A. Niño: *Fol. Chim. Theoret. Lat.* **16**, 185 (1988).
[29] Y.G. Smeyers, M.L. Senent, V. Botella, and D.C. Moule: results to be published.
[30] Y.G. Smeyers, A. Hernandez-Laguna, and P. Galera-Gomez: *An. Quim.* **76**, 67 (1980).
[31] Y.G. Smeyers and A. Hernandez-Laguna: *IntJ. Quantum Chem.* **22**, 681 (1982);
    J. Mol. Struct. (Theochem) **149**, 127 (1987).
[32] W.G. Fateley and F.A. Miller: *Spectrochim. Acta* **18**, 997 (1962).
[33] P. Groner, G.A. Guirgis, and J.R. Durig: *J. Chem. Phys.* **86**, 565 (1987).

Subject Index

TOPICS IN
MOLECULAR ORGANIZATION AND ENGINEERING

Honorary Chief Editor: W. N. Lipscomb, Harvard, U.S.A.
Executive Editor: Jean Maruani, Paris, France

1. J. Maruani (ed.): *Molecules in Physics, Chemistry, and Biology.*
   Vol. 1: General Introduction to Molecular Sciences. 1988
   ISBN 90-277-2596-9
2. J. Maruani (ed.): *Molecules in Physics, Chemistry, and Biology.*
   Vol. 2: Physical Aspects of Molecular Systems. 1988
   ISBN 90-277-2597-1
3. J. Maruani (ed.): *Molecules in Physics, Chemistry, and Biology.*
   Vol. 3: Electronic Structure and Chemical Reactivity. 1989
   ISBN 90-277-2598-5
4. J. Maruani (ed.): *Molecules in Physics, Chemistry, and Biology.*
   Vol. 4: Molecular Phenomena in Biological Sciences. 1989
   ISBN 90-277-2599-3
5. E. Schoffeniels and D. Margineanu: *Molecular Basis and Thermodynamics of*
   Bioelectrogenesis. 1990   ISBN 0-7923-0975-8
6. A. Lund and M. Shiotani (eds.): *Radical Ionic Systems.* Properties in
   Condensed Phases. 1991   ISBN 0-7923-0988-X
7. P.I. Lazarev (ed.): *Molecular Electronics.* Materials and Methods. 1991
   ISBN 0-7923-1196-5
8. E. Rizzarelli and T. Theophanides (eds.): *Chemistry and Properties of*
   Biomolecular Systems. 1991   ISBN 0-7923-1393-3
9. L.A. Montero and Y.G. Smeyers (eds.): *Trends in Applied Theoretical*
   Chemistry. 1992   ISBN 0-7923-1745-9

KLUWER ACADEMIC PUBLISHERS – DORDRECHT / BOSTON / LONDON